Myth *and* Meaning

This book is gratefully dedicated to
Susan Ward,
a highly valued personal friend and
patron of the Rock Art Research Institute.

Myth *and* Meaning

San-Bushman Folklore
in Global Context

J. D. Lewis-Williams

Left Coast Press Inc.
Walnut Creek
California

LEFT COAST PRESS, INC.
1630 North Main Street, #400
Walnut Creek, CA. 94596
http://www.LCoastPress.com

ISBN 978-1-62958-154-5 hardcover
ISBN 978-1-62958-156-9 institutional eBook
ISBN 978-1-62958-157-6 consumer eBook

Library of Congress Cataloging-in-Publication Data

Lewis-Williams, J. David, author.
Myth and meaning : San-Bushman folklore in global context / J.D. Lewis-Williams.
pages cm
Includes bibliographical references and index.
ISBN 978-1-62958-154-5 (hardback : alk. paper) — ISBN 978-1-62958-156-9 (institutional ebook) — ISBN 978-1-62958-157-6 (consumer ebook)
1. San (African people)—Folklore. 2. Mythology, San. 3. Cosmology, San. 4. Shamanism—Africa, Southern. I. Title.
GR358.2.B83L49 2015
398.208996'1—dc23
2014048986

Printed in the United States of America

The paper used in this publication meets the minimum requirements of American National Standard for Information Sciences—Permanence of Paper for Printed Library Materials, ANSI/NISO Z39.48–1992.

CONTENTS

List of Illustrations 6
About This Book 7
Acknowledgements 9
Note on Pronunciation and Spelling 11

Prologue: A Broken String? 13

Chapter 1 Myth in its San Incarnation 33

Chapter 2 Bringing Home the Honey 47

Chapter 3 The Mantis Makes an Eland 75

Chapter 4 The Fight with the Meerkats 97

Chapter 5 A Visit to the Lion's House 117

Chapter 6 The Mantis Dreams 133

Chapter 7 Narrating and Painting 149

Chapter 8 People of the Eland 173

Chapter 9 The Broken String 183

Chapter 10 'They Do Not Possess My Stories' 201

Notes 211
References 217
Index 233
About the Author 249

ILLUSTRATIONS

Figures

1.1 Map of southern Africa showing San groups 34
1.2 Wilhelm Bleek 35
1.3 Lucy Lloyd 36
1.4 Diä!kwain 38
2.1 Joseph Millerd Orpen 48
2.2 Joseph Orpen's 1874 article 54
3.1 /Han≠kass'o 79
3.2 Relationships within the Mantis's family 83
3.3 San rock painting of trance dancers and bees 94
5.1 San rock painting of human figures blending with karosses 123
7.1 San rock painting showing two men's hunting bags (conical) 156
7.2 Conflict scene 158
7.3 San rock painting of a man playing a musical bow 164
7.4 San rock painting of rain shamans leading a rain-animal 165

Section of colour plates appears after page 131.

1 The /Xam San homeland
2 The Maloti-Drakensberg
3 //Kabbo
4 Antelope skin San bags
5 Bags turning into antelope
6 Rock painting of a running bowman
7 Man playing a musical bow
8 An eland emerges from the rock face
9 Paintings made by Mapote in 1930
10 San thumb-piano
11 Ju/'hoansi people talking to Megan Biesele
12 The postapartheid South African coat of arms

ABOUT THIS BOOK

Writers on mythology face a number of dilemmas. Should they discuss the many theories of myth that have been developed over the last two centuries and then illustrate each by considering a specific myth that their readers might be expected to know? Or should they avoid any discussion of theory and simply set about presenting their own understanding of one or more myths from a particular group of people? Either way, many writers find the myths of ancient Greece looming over them. The Greek pantheon and tales about the gods' dealings with one another and with human beings have become the epitome of mythology. Other people's myths tend to be seen in the light of what we think we know about ancient Greece. Did Greek mythology play a role in moulding Western thought and values, or is Western thought still guiding our understandings of Greek myths?

In dealing with the vast array of myths worldwide—few if any communities lack a mythology—some writers try to find common denominators. Despite much ingenuity on the part of these scholars, I usually find this sort of broad, synoptic survey unsatisfactory. Primarily, my disappointment is a result of my familiarity with southern African San-Bushman mythology. For instance, I initially tried but failed to find in San myths persuasive examples of Lévi-Strauss's 'mythemes' that were not my own projections of Western values. Nor did I have much success with the psychological generalisations of Jung and Freud. Their views on myth seemed to turn the San narratives into something that appeared to me foreign to San life and thought. Perhaps other writers will have better luck.

So, not wanting to ignore the worldwide context of mythology and theory and at the same time hoping to preserve something of the essential 'Sanness' of the tales, I have adopted a different approach. I do not claim that it is a new 'theory of myth'. But I believe I am advocating a way of getting at the essentials of the ethnographic contexts of myths as a protection against seeing other people's myths in the light of our own contemporary interests and values. It is in this way that I hope to place San mythology in a global context without erasing its intrinsic character.

ACKNOWLEDGEMENTS

I am grateful to many people who assisted not only with the writing of this book but also in discussions over decades of interest in San myth and rock art. Megan Biesele was my coauthor of a 1978 article that, together with a 1976 visit with her to the Kalahari San, laid a foundation for my further thought. She continues to serve the interests of the Kalahari San in numerous practical ways. Other anthropologists with whom I have had valuable discussions include the late Lorna Marshall and the late John Marshall, Lorna's daughter Elizabeth Marshall Thomas, Richard Lee, Alan Barnard, Ed Wilmsen, Richard Katz, and the late Tony Traill, whose knowledge of San languages was second to none. I mention especially Mathias Guenther, whose book *Trancers and Tricksters* (1999) is an outstanding work of scholarship and insight. Janette Deacon, a long-time friend and pioneer researcher of the Wilhelm Bleek and Lucy Lloyd Archive, has always been most helpful. David Whitley kindly read and usefully commented on a draft of the book. I have enjoyed and benefitted from many years of friendship with him.

The director of the Rock Art Research Institute (RARI) at the University of the Witwatersrand, David Pearce, a highly valued previous coauthor, has given much assistance in the field, in protracted lunch-table discussions, and in many other ways. His current work on dating San rock is proving seminal. Other RARI members have always provided stimulating discussion and assistance. They include, especially, Mark McGranaghan, whose Oxford DPhil. thesis on the Bleek and Lloyd Archive is a remarkable piece of work, and Sam Challis, another former coauthor, whose Oxford DPhil. and subsequent publications deal with the relationship between the Maloti San and adjacent peoples. Catherine Namono and Siyakha Mguni, also members of RARI, have participated in many useful discussions. Ghilraen Laue, a RARI Ph.D. candidate, assisted with the compilation of illustrations. Eric Wettengel took photographs especially for the book. David Witelson, a Master's degree candidate, meticulously read and corrected drafts of all the chapters, thus saving me considerable embarrassment. Others who have kindly commented on drafts of individual chapters include Geoffrey Blundell, Jeremy Hollmann, Thomas Huffman, Elwyn Jenkins, Robert Leslie, Johannes Loubser, Alan Morris, and Gavin Whitelaw. Electronic scanning of rock art images was undertaken by the South African Rock Art Digital Archive (SARADA), a RARI entity under the direction of Azizo da Fonseca. All in all, it will be clear that RARI is an exceptional base for rock art research. I also thank Stacey C. Sawyer for her meticulous editing of the text and for numerous helpful suggestions.

Over many decades, the staff of the Jagger Library, University of Cape Town, has been helpful in providing access to and images from the Wilhelm Bleek and Lucy Lloyd Archive. As on numerous previous occasions, UCT has granted permission to quote from the archive and to use illustrations drawn from it. This book also contains, in reorganized and reassessed form, material that previously appeared in part in journal articles. Chapter 2: Southern African Humanities 2010, 22:1–18; Chapter 4: African Studies 1998, 26:195–216; Chapter 6: Cambridge Archaeological Journal 2013, 23: 241–62. Some material also comes from chapters in books; they are detailed in the reference list: Chapter 5 (Lewis-Williams 1996), Chapter 8 (Lewis-Williams 1998).

The royalties from this publication are donated to the Kalahari Peoples' Fund.

J.D. Lewis-Williams
March 2015

NOTE ON PRONUNCIATION AND SPELLING

In addition to the more usual phonetic representations, the following symbols are used for the clicks that are a distinctive feature of the Khoisan language family. I take the descriptions of these sounds from Lorna Marshall's book The !Kung of Nyae Nyae (Marshall 1976:xx).

/ Dental click The tip of the tongue is placed against the back of the upper front teeth; in the release, it is pulled away with a fricative sound. English-speakers use a similar sound in gentle reproof.

! Alveolar-palatal click The tip of the tongue is pressed firmly against the back of the alveolar ridge, where it meets the hard palate and is very sharply snapped down. A loud pop results. English-speakers use this sound to imitate horses' hoofs on paving.

≠ Alveolar click The front part of the tongue, more than the tip, is pressed against the alveolar ridge and drawn sharply downward when released.

// Lateral click The tongue is placed as for the alveolar click. It is released at the sides by being drawn in from the front teeth. Horse riders sometimes use lateral clicks to signal their steeds to start or go faster.

ʘ Labial click The frontal closure is made with pursed lips; when the lips are released, the sound is like a kiss. This click is found in southern San languages only.

X In San orthography, X indicates a guttural sound as in the Scottish loch.

In Bantu language (for example, isiZulu, isiXhosa) orthography, clicks, which derive from Khoisan languages, are represented as follows:

/ = c
! = q
// = x

Contrary to strict phonetic practice, I have given the first alphabetical letter of a proper noun as a capital (for instance, //Kabbo). This departure makes for easier reading by persons not familiar with San names.

Spelling

There have been many spellings of San words used by different linguists, anthropologists, and even San groups themselves over the years. I retain the spellings as they are found in the literature, even though they may not be consistent over the course of the book. The most recent orthography, especially the Ju/'hoan one that Patrick Dickens compiled and that the people themselves and the Namibian government now adopt, has become one of the several educational languages of Namibia.

PROLOGUE

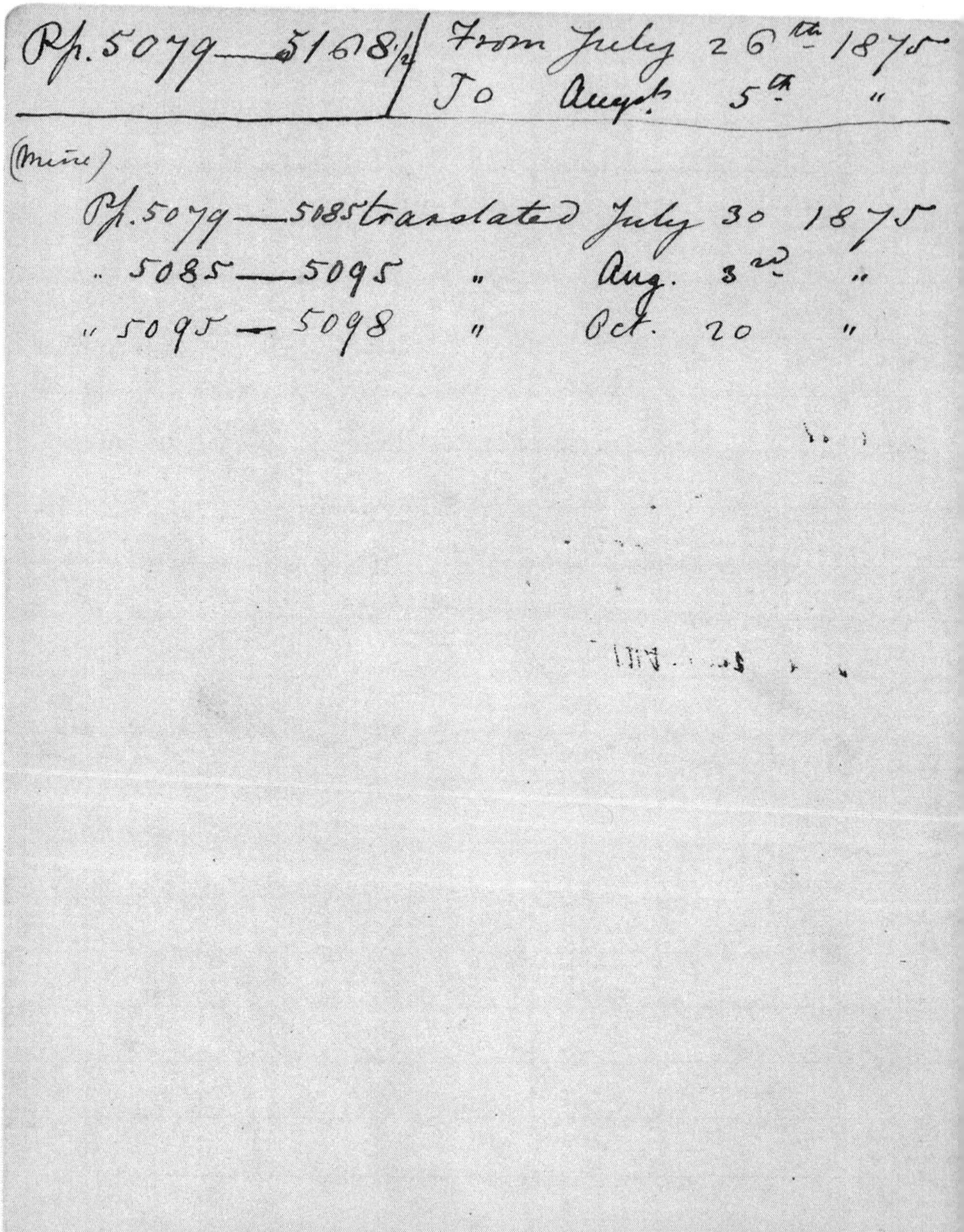

Pp. 5079—5168½ | From July 26th 1875
To Augst. 5th "

(Mine)

Pp. 5079—5085 translated July 30 1875
" 5085—5095 " Aug. 3rd "
" 5095—5098 " Oct. 20 "

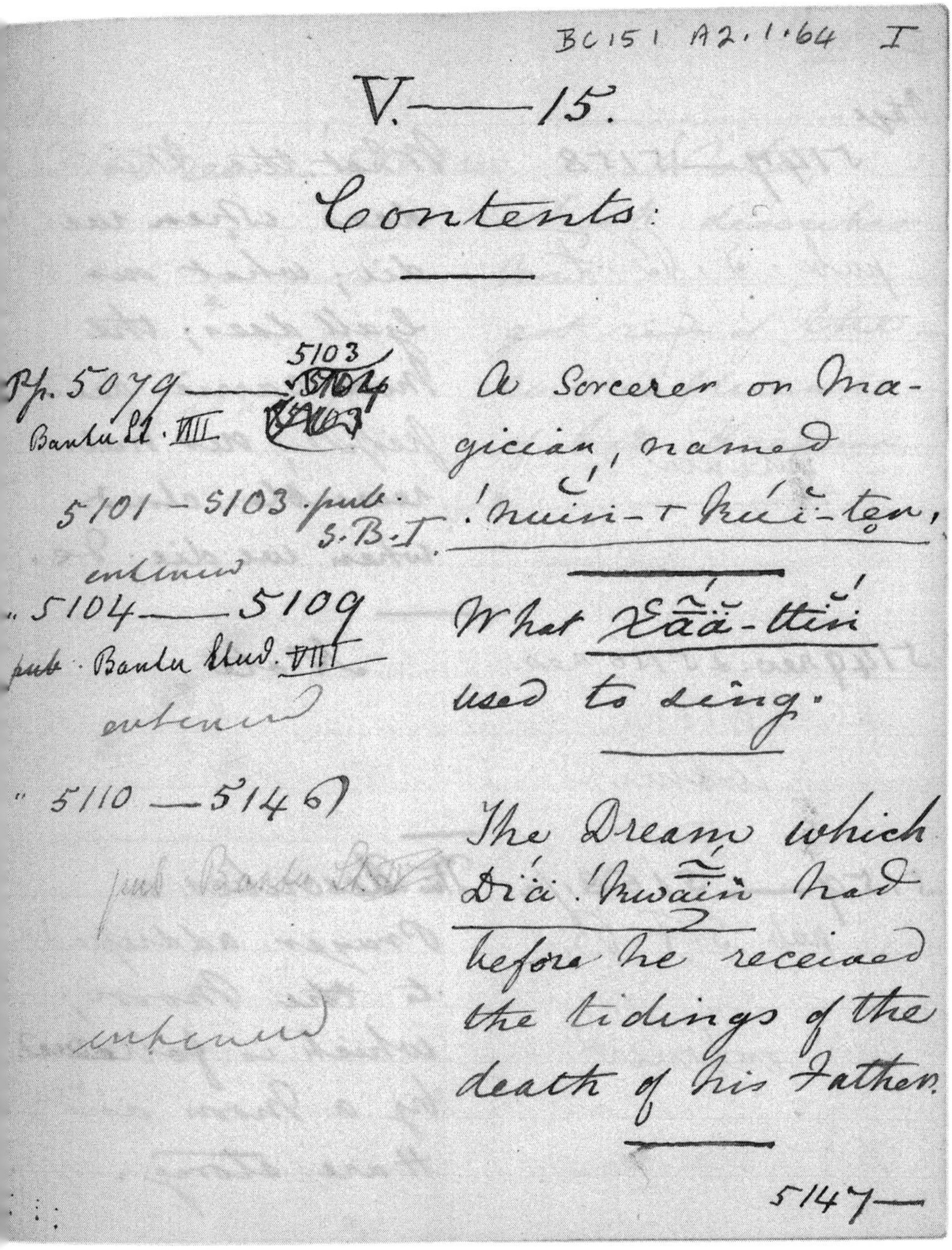

BC151 A2.1.64 I

V.—15.

Contents:

Pp. 5079 —— 5103
Bantu St. VIII

5101 – 5103 pub. S.B.T.
entered

5104 —— 5109
pub. Bantu Stud. VIII
entered

5110 — 5146
pub Bantu St.
entered

A Sorcerer or Magician, named !nuin-/kúi-tẽn,

What Xãä-ttin used to sing.

The Dream which Dïä!kwain had before he received the tidings of the death of his Father.

5147—

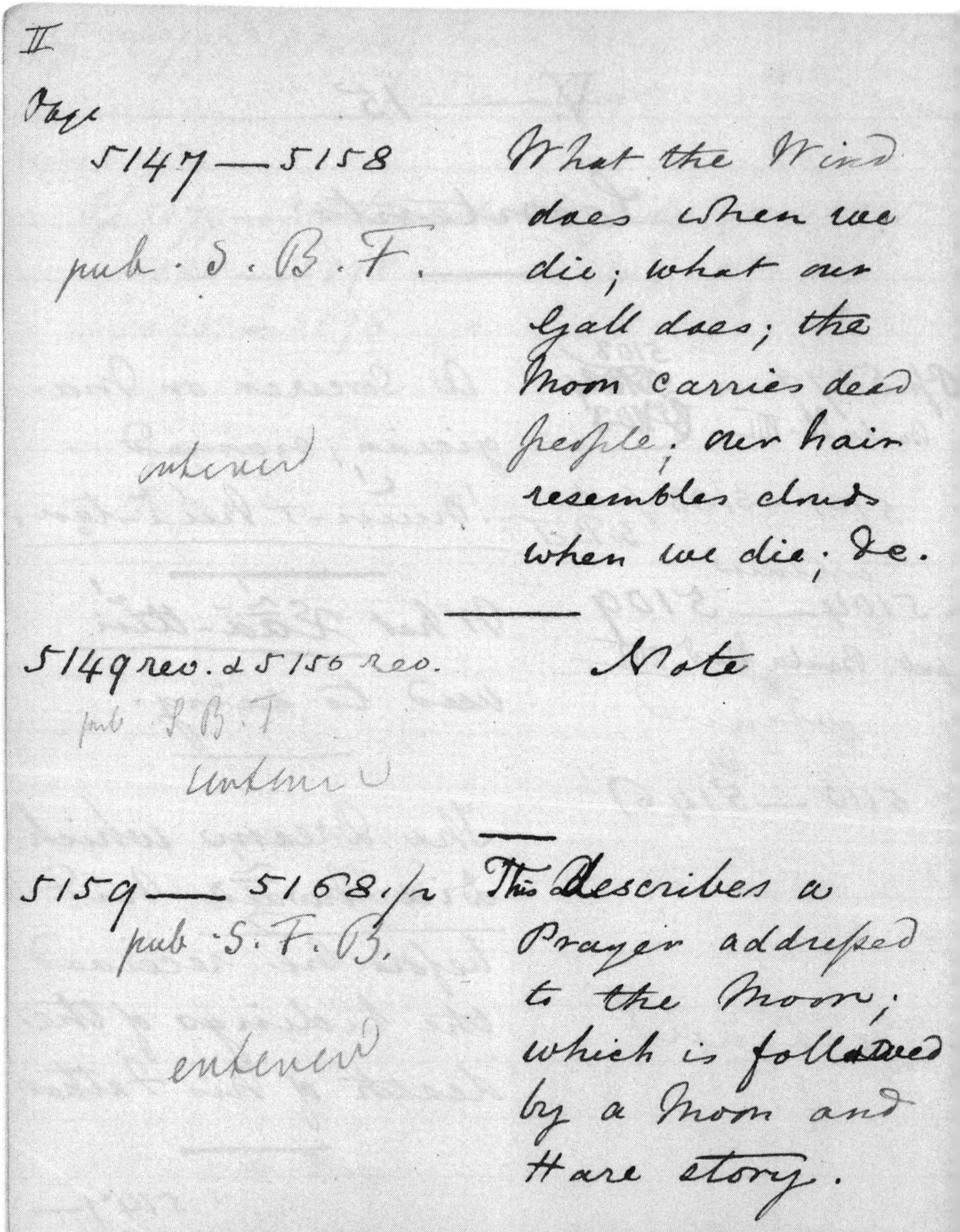

II

Page

5147—5158
pub. S. B. F.

What the Wind does when we die; what our Gall does; the Moon carries dead people; our hair resembles clouds when we die; &c.

entered

———

5149 rev. & 5150 rev.
pub S.B.F

Note

entered

———

5159—5168. /r
pub. S. F. B.

This describes a Prayer addressed to the Moon; which is followed by a Moon and Hare story.

entered

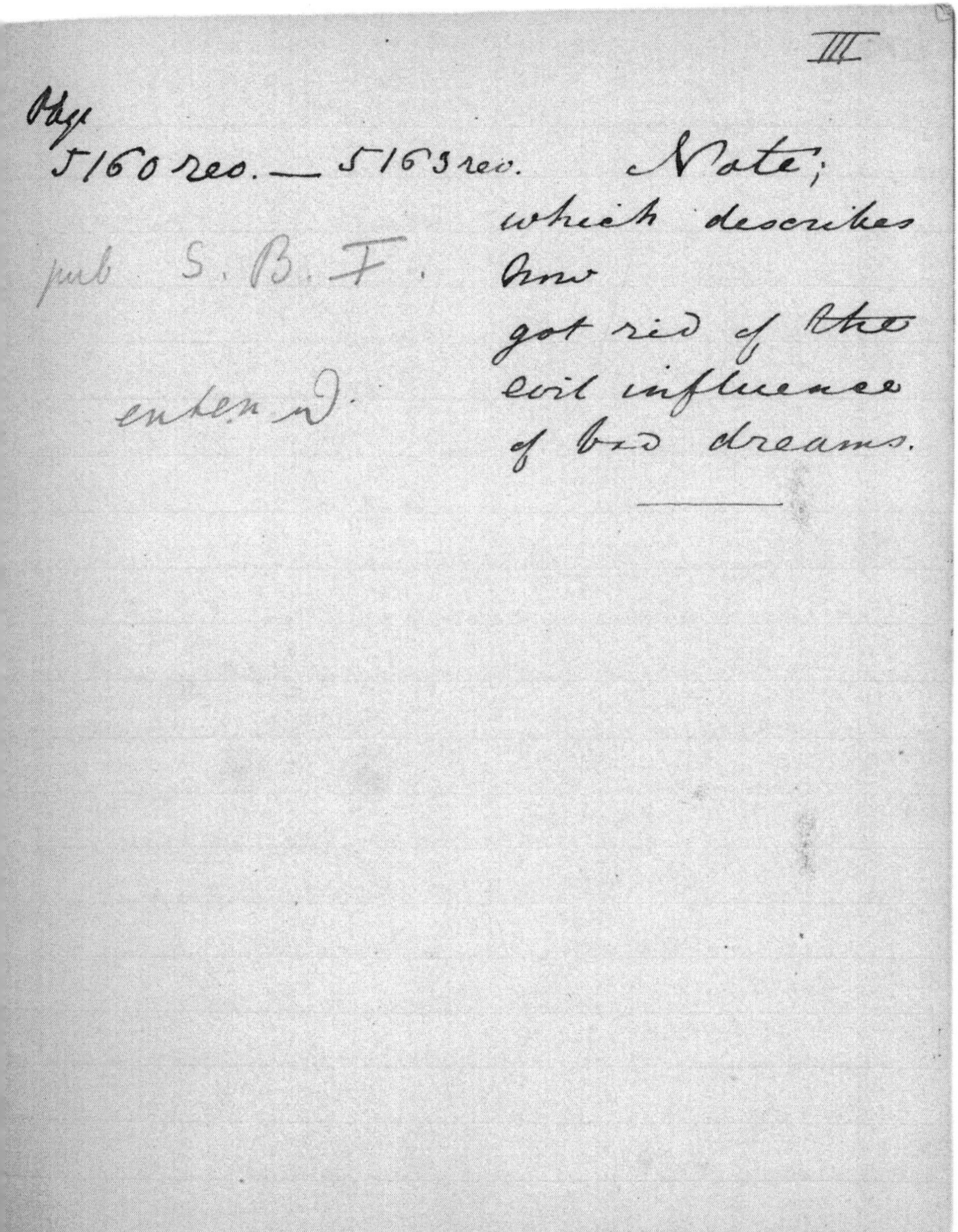

III

Phy

5160 rev. — 5163 rev. Note;

which describes

how

got rid of the

evil influence

of bad dreams.

pub S. B. F.

entered.

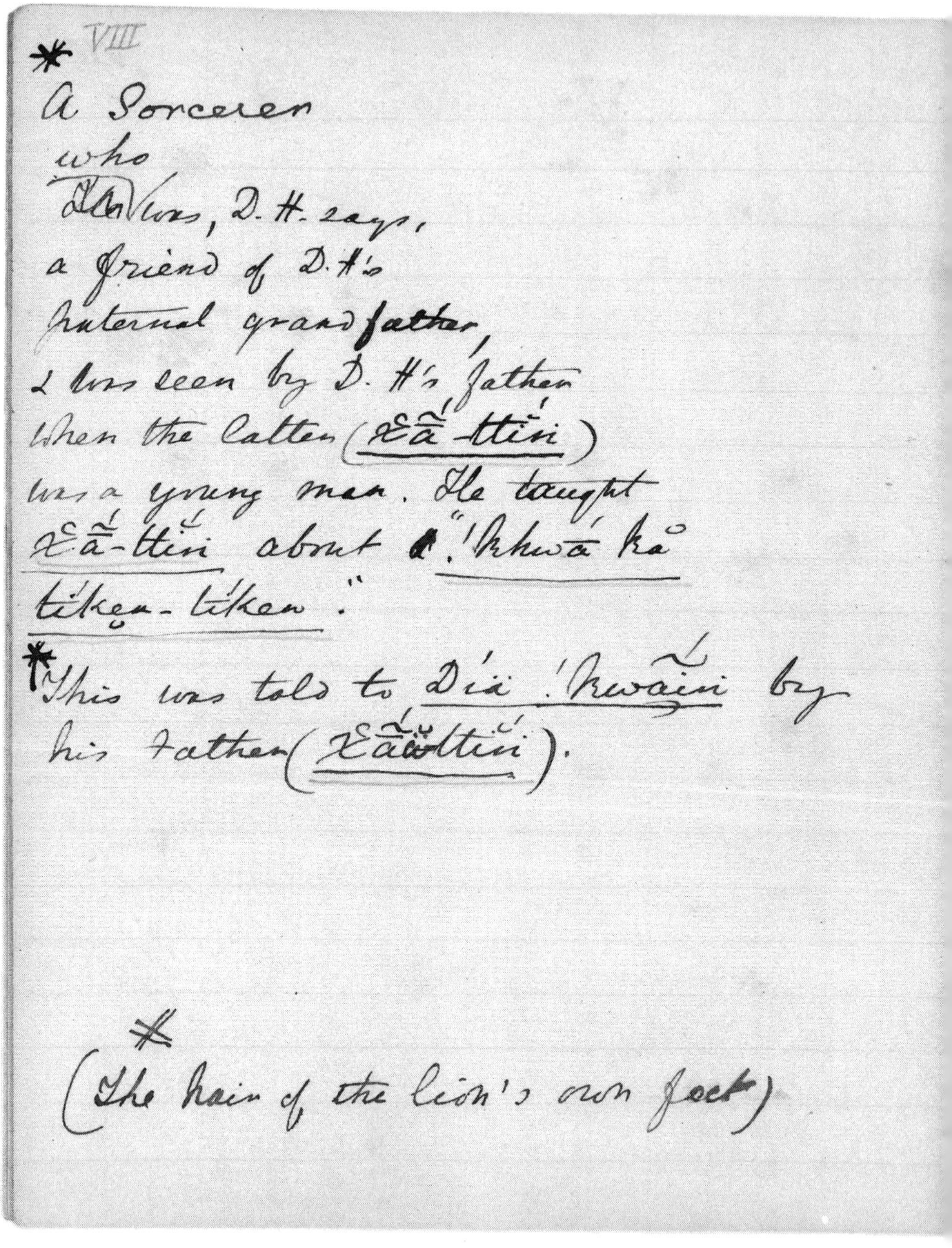
* VIII
A Sorcerer
who
~~Dā~~ Vers, D. H. says,
a friend of D. H's
paternal grandfather,
& was seen by D. H's father
when the latter (Xā-ttin)
was a young man. He taught
Xā-ttin about "!khwā ka
tiken-tiken".
*This was told to Dia !kwain by
his father (Xā-ttin).

*
(The hair of the lion's own feet)

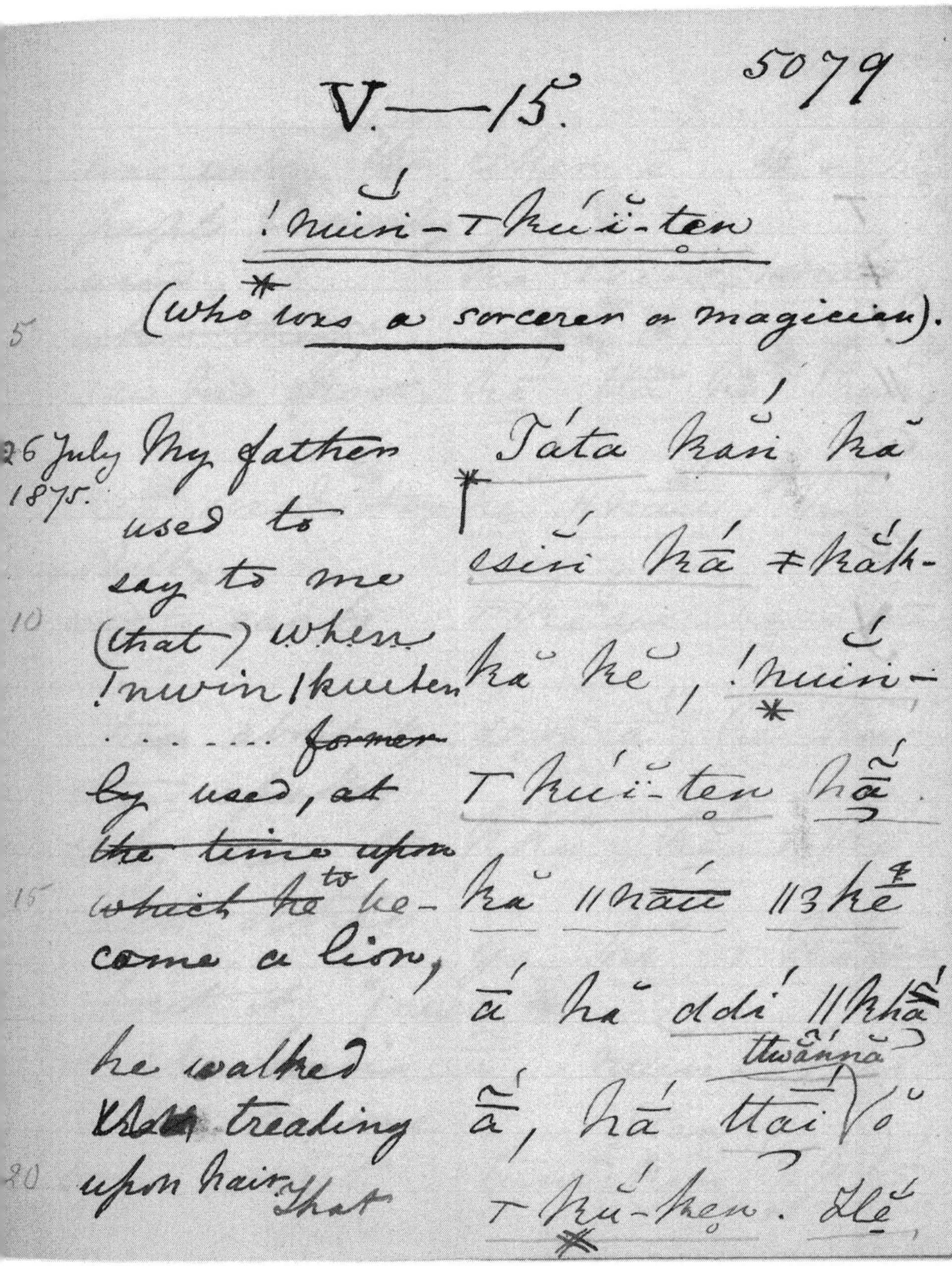
V.—15. 5079

!nuin-/kúi-ten

(who was a sorcerer or magician).

26 July 1875. My father used to say to me (that) when !nuin|kuiten ...... formerly used, at the time upon which he to become a lion, he walked treading upon hair. That

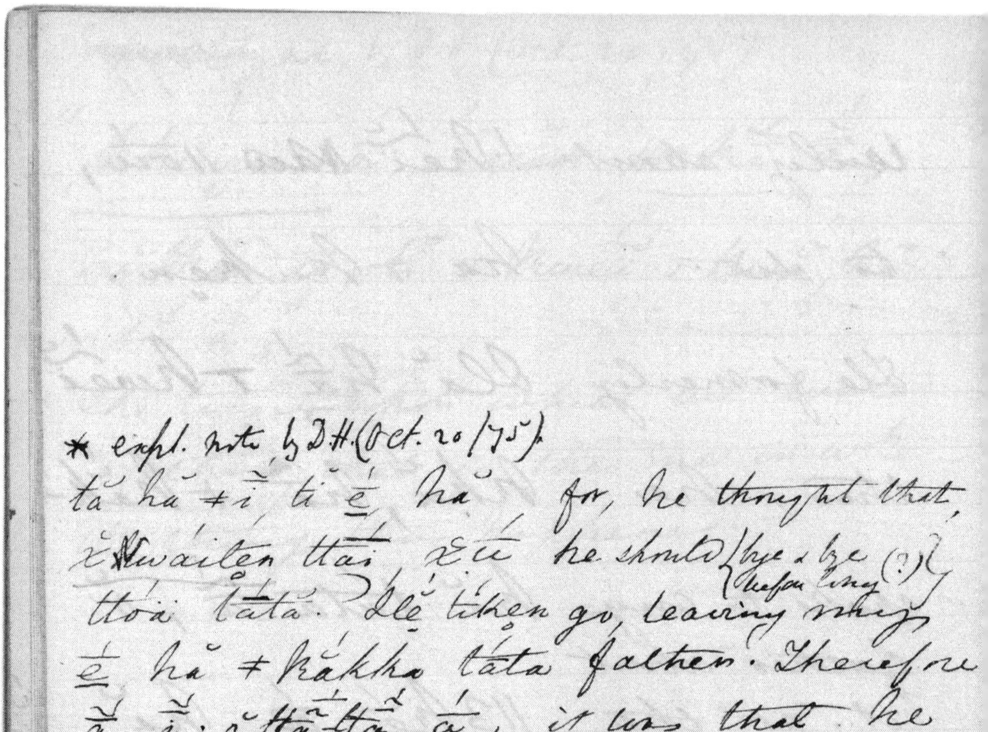

* expl. note by D.H. (Oct. 20/75).

for, he thought that, he should {bye & bye (?) / before long} go, leaving my father. Therefore it was that he spoke to my father about it, ... Things felt as if he must soon go & leave him

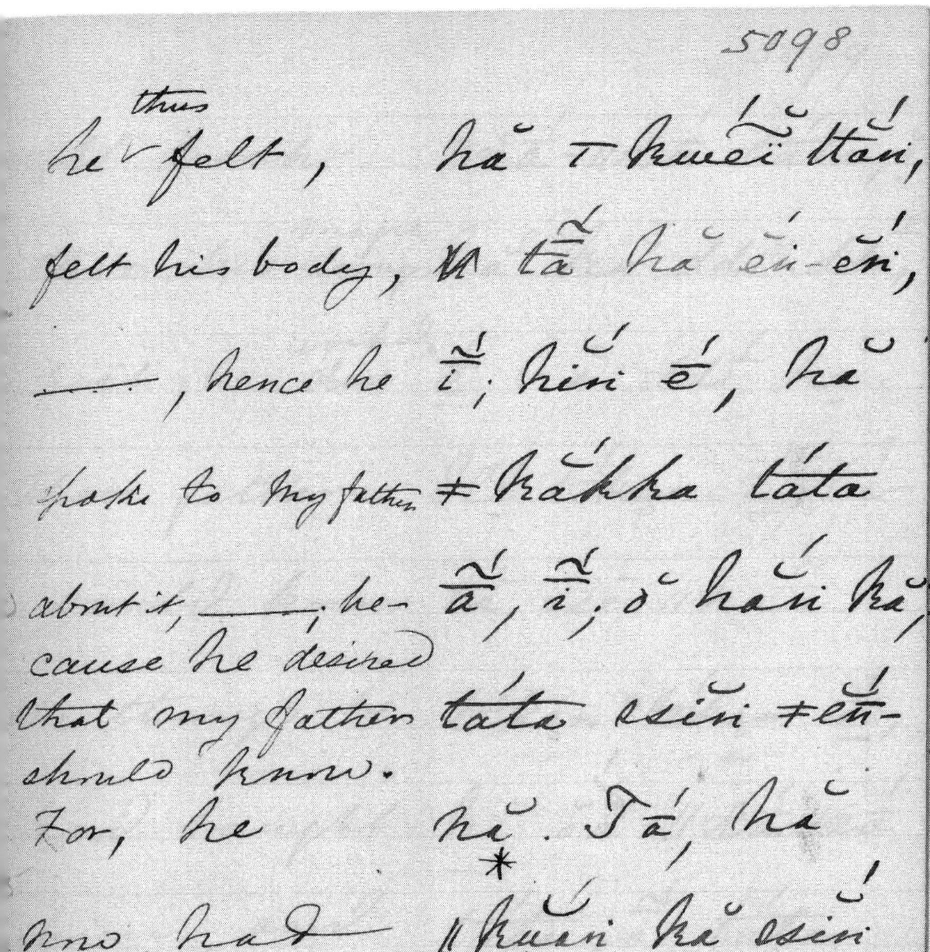
5098

he thus felt, hă ⊤kwēï ttan,

felt his body, ha tā̃ hă ēi-ēri,

—, hence he ĩ; hĭri ē, hă

spake to my father ≠kăkka tāta

about it, —, be- ā̃, ĩ; ō hări hă;
cause he desired

that my father tāta ssiri ≠ēn-
should know.

For, he hă. Tā, hă
*

who had ||kŭäri hă ssin

desired to hă hă !khāu tá-
take my father

with him, that tă, hă sse
he

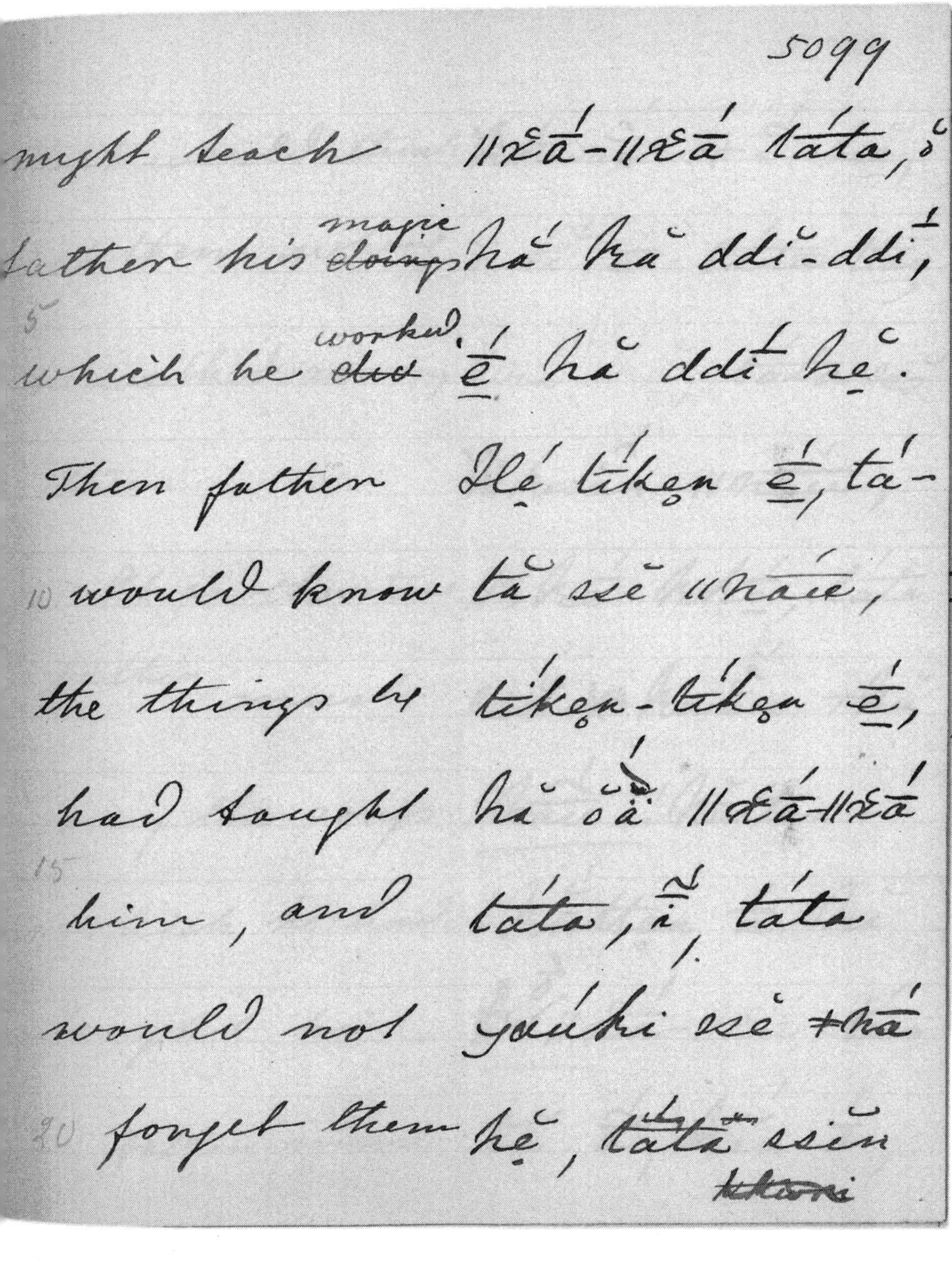
5099

might teach ||xā-||xā tāta, ī

father his ~~doings~~ magic hā ha ddi-ddi,

which he ~~did~~ worked ē hā ddi hē.

Then father Ikē tiken ē, ta-

would know tā sse ||nau,

the things he tiken-tiken ē,

had taught hā oä ||xā-||xā

him, and tāta, ī, tāta

would not yauki sse ǂnā

forget them hē, tāta ssin

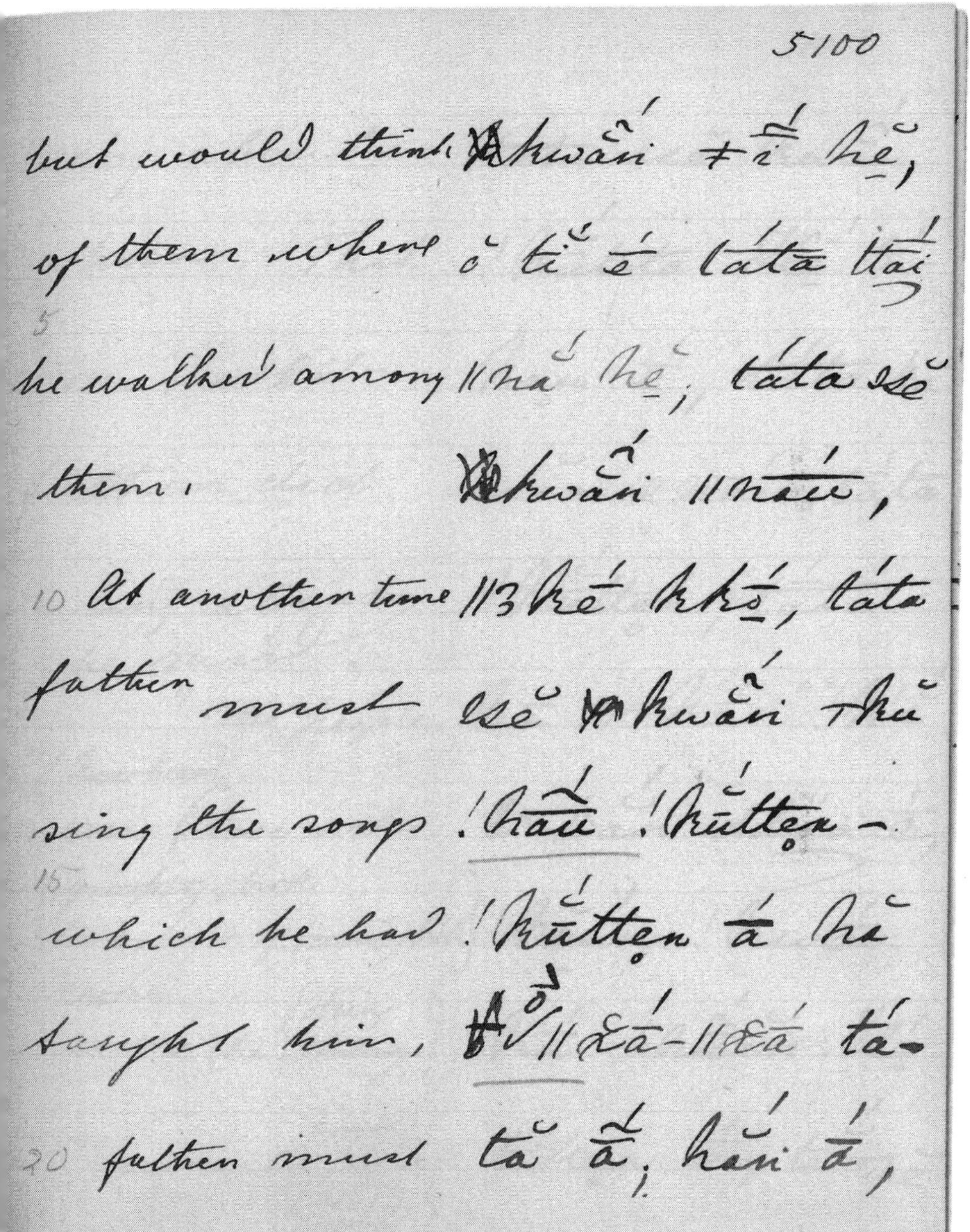

5100

but would think ǂkwăĩ ǂ ĭ hē;
of them where ŏ tĭ ē tátā ttai;
he walked among ||nă hē; táta ẽsẽ
them. ǂkwăĩ ||nau,
At another time ||ʒkē kkō, táta
father must ẽsẽ ǂkwăĩ [illegible]
sing the songs !kau !kŭttẹn-
which he had !kŭttẹn ā hă
taught him, [illegible] ||ʒā-||ʒā tá-
father must tă ā; hăĩ ā,

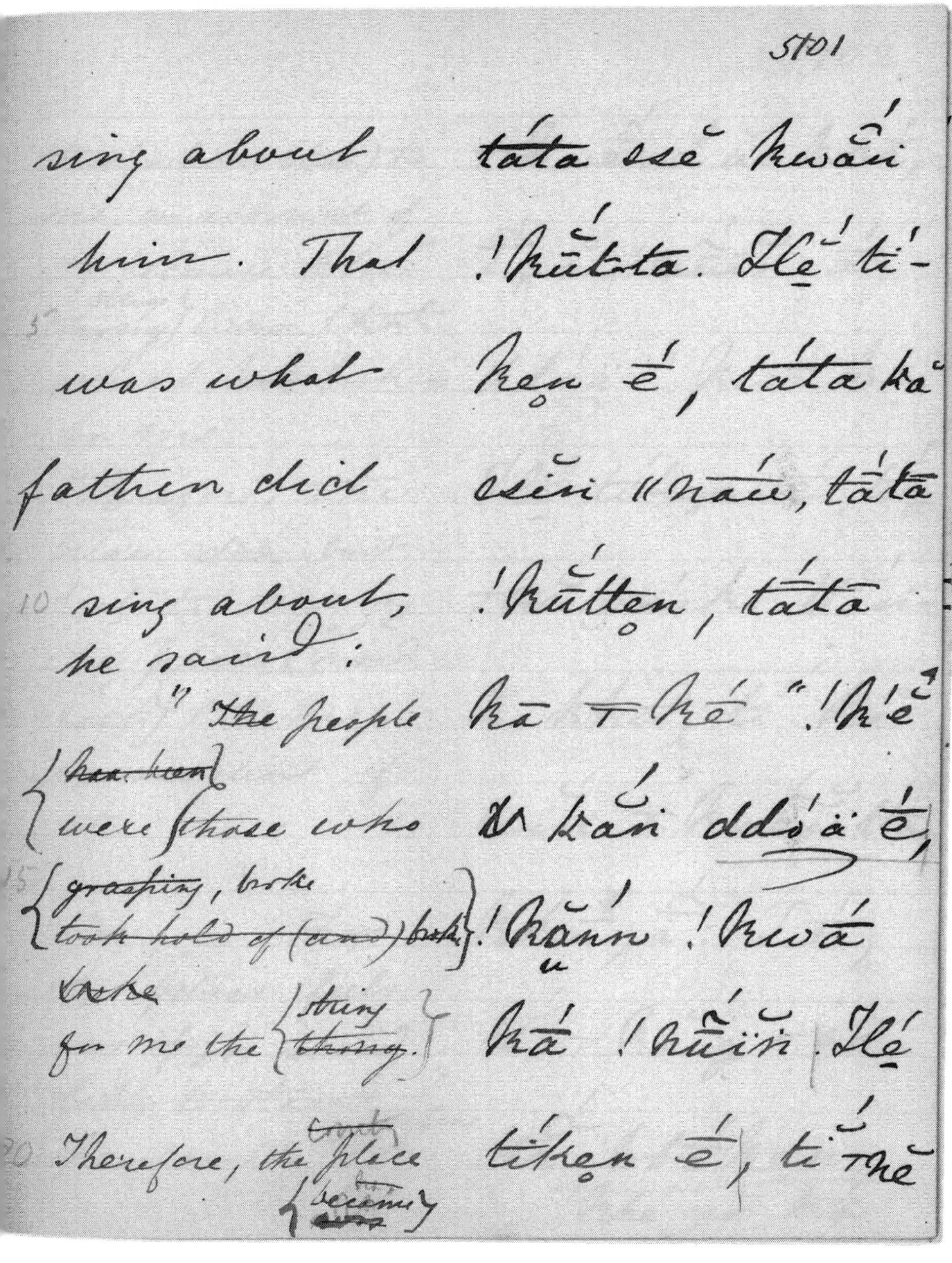

5101

sing about him. That was what father did sing about, he said: " ~~The~~ people ~~have been~~ were those who ~~grasping~~, broke ~~took hold of (and) broke~~ ~~broke~~ for me the string ~~thing~~. Therefore, the place ~~was~~ became

tata sse kwãŭ !kŭtta. Ilẹ ti-kẹn ē, tata ka osĭn «naú, tata !kŭttẹn, tata ka ‡ kē " !kē ≠ kani ddóä ē, !kann !kwá ka !kŭïn. Ilẹ tikẹn ē, ti ‡ nē

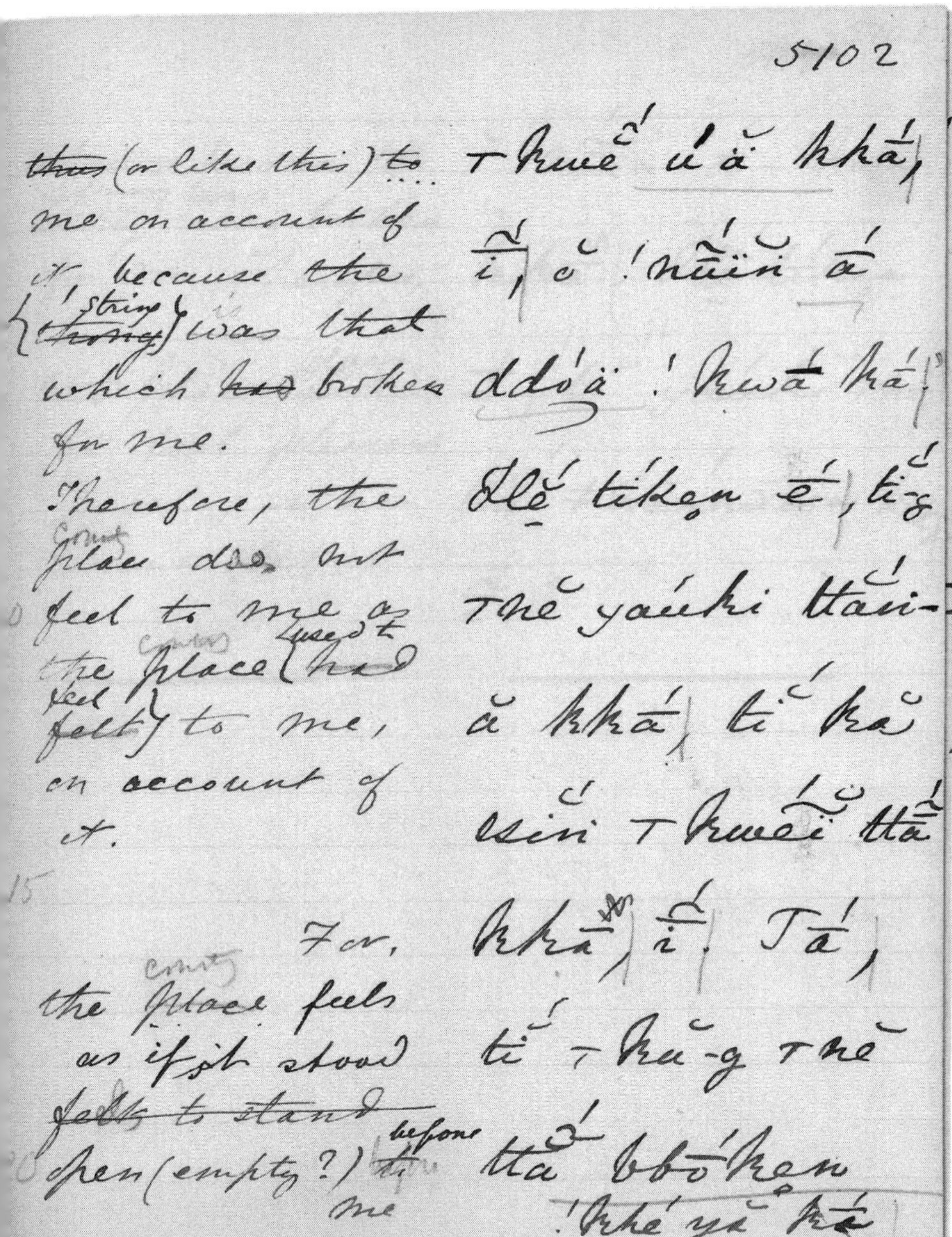

5102

(or like this) to me on account of it, because the 'string' was that which broken for me.

Therefore, the country does not feel to me as the country used to feel to me, on account of it.

For, the country feels as if it stood open (empty?) before me

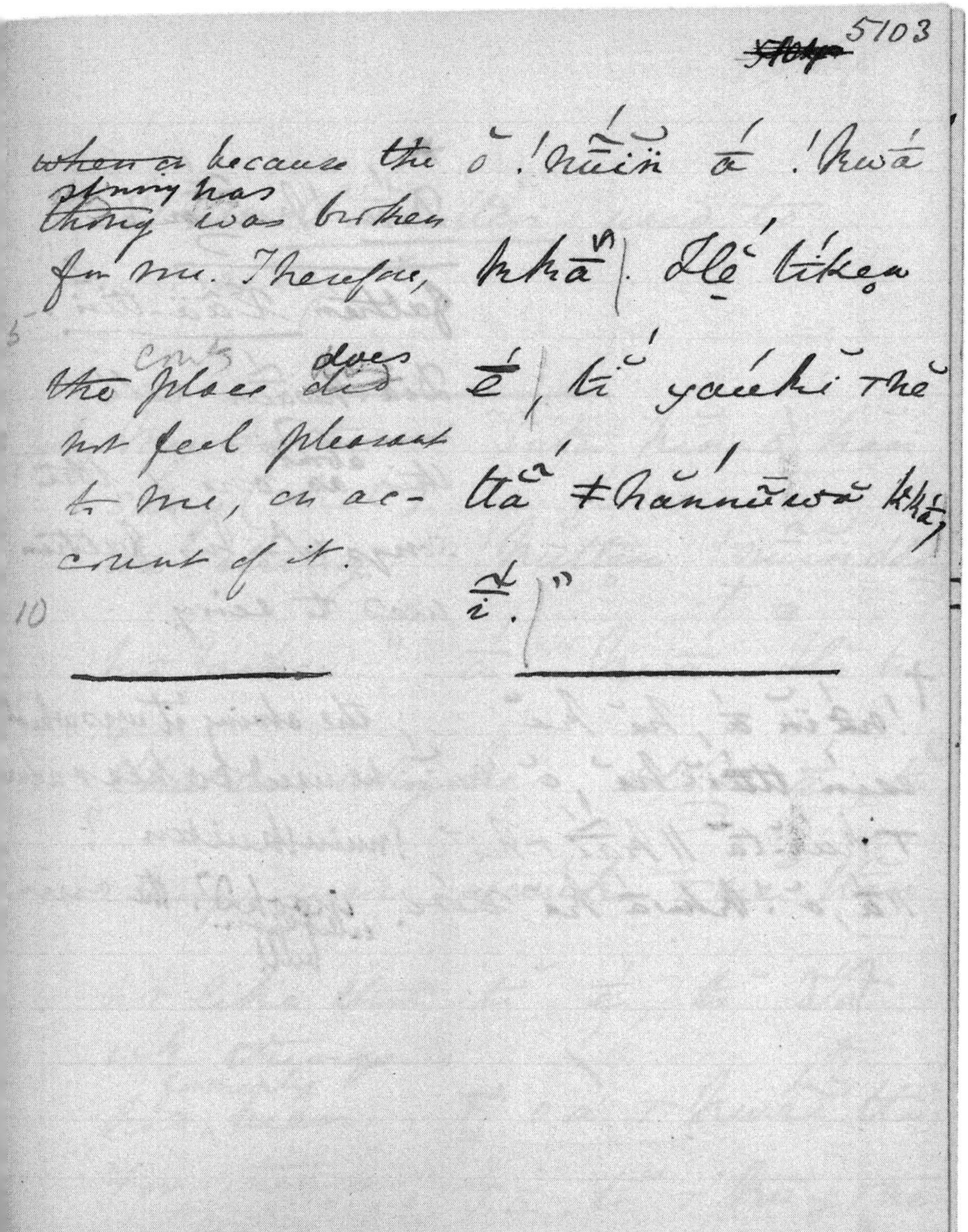
5103

because the thing has broken for me. Therefore, the place does not feel pleasant to me, on account of it.

A BROKEN STRING

Dorothea Bleek's published version of The Song of the Broken String (5101–5103)

People were those,
Who broke for me the string.
Therefore,
The place became like this to me,
On account of it,
Because the string was that which
broke for me.
Therefore,
The place does not feel to me,
As the place used to feel to me,
On account of it.
For,
The place feels as if it stood open
before me.
Because the string has broken for
me.
Therefore,
The place does not feel pleasant
to me.
On account of it.

(Bleek 1936:134; Hollmann 2004:279–283)

CHAPTER ONE

Myth in Its San Incarnation

A story is the wind. It does float along to another place.
(//Kabbo, L.II.32.2887)

On July 27, 1875, Diä!kwain, a southern African /Xam San (Bushman) man spoke in what is now an extinct language about his beliefs and experiences.[1] The facsimile manuscript pages reproduced in the Prologue to this book are part of the original record of that occasion. Both before and after that day, Diä!kwain and other /Xam people were asked to recount their myths—complex stories that, we now know, encapsulate the multifaceted essence of being a San person. Some of those tales feature in this book. Working from the narrators' actual /Xam words and their often hidden connotations, we are able to see past the surface of the stories and into less obvious, but none the less fundamental, realms of San belief and life.

But there is more at stake. Diä!kwain provided texts that over a century and a quarter later are opening up insights into the ways in which not only his but also many other people's myths around the world 'work'. Through his tales, we can begin to understand why myths in general so powerfully affect the minds and social lives of those who relate and hear them. The San narratives embody characteristics of myth that we encounter far beyond the confines of southern Africa.

To identify those features we need to understand exactly how the nineteenth-century San texts were written down; we can then assess the extent to which we can trust them as accurate records. Indeed, the processes by which the /Xam language and myths were preserved were by no means incidental or perfunctory: as we shall see, they were systematic and meticulous. Those who mastered the language and then wrote down the myths and translated them

Myth and Meaning: San-Bushman Folklore in Global Context by J. D. Lewis-Williams, 33–46

were, I believe, performing a task comparable with the much more famous decipherment of Maya glyphs and Minoan script.

Saving a Language

In the 1870s, there were many San languages, as indeed there still are among the groups who continue to live in the Kalahari Desert to the north of where the nineteenth-century /Xam lived (Figure 1.1).[2] The southern /Xam language is no longer spoken, but the 1870s work of the German philologist Wilhelm Bleek (1827–1875) and his sister-in-law and coworker Lucy Lloyd (1834–1914) managed to preserve much of it (Figures 1.2 and 1.3). It was Lloyd who recorded the pages shown in facsimile. As a young man who had been educated at the University of Bonn, Bleek himself was initiated into the intricacies of extinct languages by Karl Richard Lepsius, the celebrated Berlin philologist and Egyptologist. Bleek adapted Lepsius's orthography so that he could record the clicks and other complex sounds of San languages, sounds that are difficult for

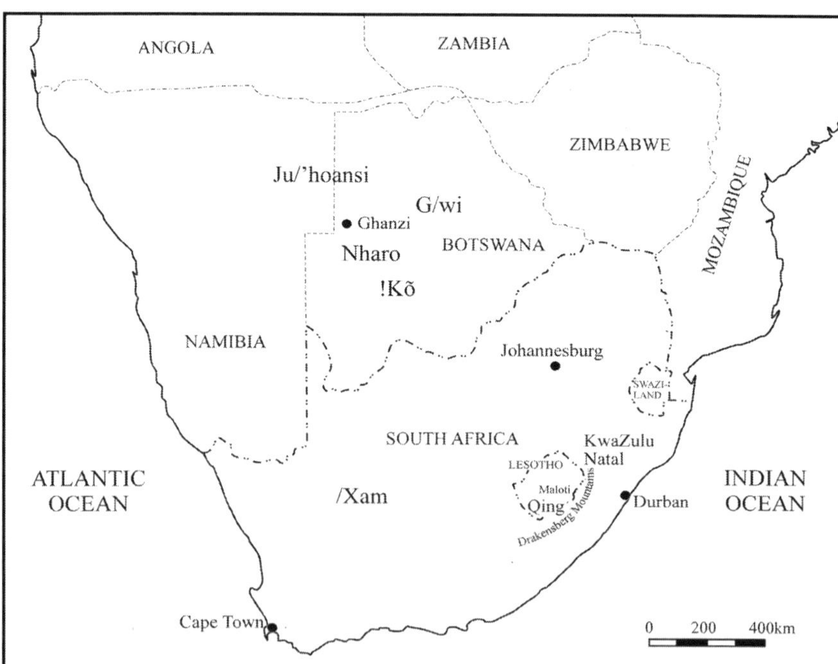

FIGURE 1.1 Map of southern Africa showing San groups

FIGURE 1.2 Wilhelm Bleek

many Westerners to pronounce. Some San languages have four clicks and some (including /Xam) have five. Bleek kept in touch with Lepsius and, years later, sent him a portrait of one of his /Xam 'teachers'—Bleek's own word (Colour Plate 3). Lepsius was delighted to have it (Bank 2006:186–87), as well he might have been had he known that Bleek and Lloyd's work would have a major impact a century and a quarter later: the new, post-apartheid South African national motto would be expressed in the extinct /Xam language: *!Ke e: /xarra //ke*—'People who are different come together' (Colour Plate 12).[3] The ancient language was resurrected and now lives on in new ways in a new political landscape. Today, descendants of the people who became known as Khoesan are a powerful element in southern African society and culture.[4]

After Bleek's death in 1875, Lloyd continued with their work and in 1911 published selected passages in *Specimens of Bushman Folklore* (Bleek & Lloyd 1911). It was through this substantial book that the world began to learn about San belief and mythology, although it tended to be overshadowed by

FIGURE 1.3 Lucy Lloyd

Sir Walter Spencer and Francis Gillen's Australian ethnographic reports and by Sir James Fraser's influential *Golden Bough: A Study in Magic and Religion*. Later, in the 1920s and 1930s, Bleek's daughter, Dorothea (1873–1948), prepared the manuscript pages shown in the Prologue together with other texts for publication, but, as the facsimile shows, the original 1870s manuscript transcriptions, with all their valuable alterations and notes, still exist. They are a seemingly inexhaustible resource. I draw on them repeatedly in the following chapters.[5] The passage given in facsimile is not a full, narrative myth, but it usefully illustrates some of the extra-narrative principles that I develop in this book. The stories of myths are of course important, but they do not tell 'the whole story'. As we shall see, there are elements in the reproduced passage and its 'Song of the Broken String' that are integral to the essence of San myth. They give us some idea of how the tales, over and above their plots, communicated to their original listeners.

The notebook cover reproduced in the Prologue begins to show how systematically Bleek and Lloyd set about their task of recording the /Xam language and lore. It indicates, in addition to its date, that it contains pages

5079 to 5168½. The facsimile pages that follow the cover comprise the introductory index to the notebook, the page numbered 5079 (the start of an account of a 'sorcerer or magician' named !Nuin-/kúiten, whom I discuss in later chapters), and then pages 5098 to 5103. The final page number of this notebook, 5168½, may seem to indicate a lot of pages overall, but, altogether, the Bleek and Lloyd Archive comprises as many as 12,000 pages. They are numbered on the right only. The left-hand pages were designated verso (indicted by an apostrophe) and reserved largely for notes and explanations. For instance, facsimile page 5097' contains an explanation that was added on October, 20 1875. As the facsimiles show, each of the right-hand pages has a column in which Bleek or Lloyd recorded the /Xam language in phonetic script, while the other column has an English transliteration. The phonetic /Xam columns shown in facsimile are comparatively free from alterations, whereas the English columns have changes and signs of indecision. It seems that Lloyd must have made an initial phonetic record that is now lost before writing out the fair copy in the notebook and then proceeding to the task of translation.

Decades after the texts were recorded, Dorothea pencilled annotations on the three index pages of the notebook to indicate which passages Lloyd had selected for publication in *Specimens of Bushman Folklore* and which she herself published in the 1930s in the journal *Bantu Studies*. Most of the archive, however, remains unpublished, although a complete electronic scan is available.[6] In indexes of the whole collection, although not on the covers, each notebook number is prefaced with a B or an L to indicate whether it was Bleek's or Lloyd's. The large V on the cover shown in facsimile signifies the /Xam informant: Lloyd allocated each of them a Roman numeral.[7] The following Arabic number shows that it is this informant's fifteenth notebook. The parenthetical initials D. H. stand for the informant's acquired European name: David Hoesar. It may derive from the name of a colonist's farm, today known as Hoezar Wes, where he may have worked for a period (Deacon 1996a:32). His San name was Diä!kwain (Figure 1.4).

It will be readily apparent that the circumstances in which Bleek and Lloyd made their record were very different from those in which the narrators told the tales in their own cultural circumstances: they would have dramatised and varied them, as those San living in the Kalahari still do. Nevertheless, it is important to remember that the texts in the Bleek and Lloyd Archive are not homogenised, anonymous summaries contrived by Westerners. We know not only the names of the speakers but also something of their lives.

FIGURE 1.4 Nineteenth-century photograph of Diä!kwain

The /Xam San

It is, of course, within their own (inevitably changing) historical contexts that all myths, not just those of the San, should be understood. We therefore need to ask: what sort of lives were Diä!kwain and the other /Xam people living in the 1870s?

For many thousands of years, their ancestors were traditional hunters and gatherers on the semi-arid plains of what is now the Northern Cape Province of South Africa (Colour Plate 1). They kept neither flocks nor herds. Traditionally, they had no chiefs or politically constituted 'tribes'. In more recent times the /Xam lived their foraging lives adjacent to Khoekhoe pastoralists. Farther to the east, linguistically related San groups lived alongside Bantu-speaking agriculturalists, who also kept cattle (Chapter 2). The major disruption to the /Xam's lives was, however, caused by European colonists who began to settle in the region in the late eighteenth and early nineteenth centuries. The /Xam were not driven north into the Kalahari Desert by invading 'stronger races', as is commonly believed. Linguistically distinct San groups and their ancestors have lived in the Kalahari for thousands of years.[8] Unlike the northern San, the southern San were exterminated or to some extent absorbed into other populations.[9] The Bleek and Lloyd notebooks remain our richest record of their lives and beliefs.

Diä!kwain's life gives a broad, introductory idea of how /Xam people were living on the Cape frontier in the 1870s (Deacon 2001). He came from a large family and frequently spoke of his siblings and other relatives (Deacon 1996a). His father died when he was a teenager. His mother had simply disappeared; he believed she had been killed by the Boers, immigrant farmers of Dutch descent.[10] His first wife died of an illness in about 1863, and his paternal grandmother died when he was a youth. Her name, /Xarraŋ/xarraŋ, means that she was different from ordinary people.[11] She was believed to have supernatural powers and was said to have a head and gait like those of an ostrich. A rainmaker by the name of /Kăŭnŭ was also said to have eyes like an ostrich (Bleek 1933:390), so /Xarraŋ/xarraŋ was probably a 'medicine person'. Even after her death, if people uttered her name irreverently, the wind blew and 'the place became unpleasant'. Diä!kwain himself seems to have had some experience of /Xam weather-altering practices, as the 'Song of the Broken String' and other passages imply. His father, Xa-ttin, was a rainmaker. Indeed, his family was closely associated with what Bleek and Lloyd called 'sorcery', and he had much to say on the subject—as I do in this book.

In an echo of more peaceful times, Diä!kwain said that Xa-ttin had made rock engravings of 'gemsbok, quaggas, ostriches, etc.' at a place called !Kann, where animals had come to drink, as he ominously put it, 'before the coming of the Boers' (Bleek & Lloyd 1911:xiv). Using a very rough sketch map that Bleek drew on instructions given to him by one of the /Xam people (distances are shown as days' journeys), the archaeologist Janette Deacon (1986, 1988) tracked down the original homes of Bleek and Lloyd's /Xam teachers—a considerable feat. !Kann itself is not marked on Bleek's map, but Deacon found rock engravings at a natural spring about 12 km from the place where Xa-ttin was said to have lived (Deacon 1986:147). Diä!kwain was the only /Xam informant with a direct link to a maker of rock art (Deacon 1996a:35). Nevertheless, the Bleek family's /Xam people spoke a lot about copies of rock paintings that they were shown (rather than those of rock engravings that they were also shown; the San employed both techniques). A crucial question is how—if at all—the myths that Diä!kwain and other /Xam people recounted related to the images. I deal with this problem in Chapter 7, after I have examined specific myths.

Years later and drawing on family tradition, Dorothea Bleek and her sister Edith described Diä!kwain as 'a soft-hearted mortal, who would not, unprovoked, have hurt a fly' (Bleek & Bleek 1909:40). Yet he had been found guilty of having shot and killed a Boer farmer. He was incarcerated in the Cape Town Breakwater Prison, where Wilhelm Bleek, seeking San speakers, first encountered him. The judge who sentenced him in 1869 to a comparatively light five-year term of penal servitude may have been impressed by his quiet nature and perhaps recognised that he had acted in self-defence (Deacon 1996a:34). It seems that the farmer had threatened to fetch his gun and shoot Diä!kwain and his whole family—not unusual behaviour on the lawless frontier.[12] Wilhelm Bleek eventually managed to persuade the British governor of the Cape Colony to allow Diä!kwain and other San prisoners to live with him in his suburban home so that he could study their language and lore. While with the Bleek family, Diä!kwain escorted their lady friends home at night; he offered to protect them on their somewhat lonely way and became known as Dr. Bleek's 'pet murderer' (Bleek & Bleek 1909:40). When his penal term expired, he chose to remain with the Bleeks for a while before returning to the area where he had lived on the colonial frontier. The purpose of this long and difficult journey was to fetch his sister and brother-in-law; he came back with them on June 13, 1874. When Bleek died in 1875, Diä!kwain again decided to remain with the family so that he could continue to assist Lucy Lloyd with her San researches. There was, in any event, probably little left for him on the devastated frontier. Indeed, it is suspected that, on his final return home in March 1876, friends of the farmer he had shot, angered by his lenient sentence, killed him (Deacon 1996a:34).

Relating to the San

How did the Bleek family, living in a Victorian colonial *milieu,* respond to people like Diä!kwain? There is no simple answer to this question.[13] As their understanding of the San developed, they moved from seeing them as objects of philological study to a more humane view. Lloyd's considerable affection for her teachers is frequently evident. Decades later, the titles that Dorothea Bleek gave to the nine collations of texts that she chose for publication in *Bantu Studies* added up to her own personal image of the San and the sort of lives they lived: 'Baboons', 'The Lion', 'Game Animals', 'Omens, Wind-Making, Clouds', 'Rain', 'Rain-Making', 'Sorcerers', 'More about Sorcerers and Charms', and 'Special Speech of Animals and Moon Used by the /Xam Bushmen' (Bleek 1931–1933, 1935a, 1936). She seems to have regarded the /Xam as simple hunting people with some rather strange but intriguing beliefs. Even so, her summary view of them is surprising: 'The Bushman is a good lover and a good hater, very loyal and very revengeful. He remains all his life a child, averse to work, fond of play, of painting, singing, dancing, dressing up and acting, above all things fond of hearing and telling stories' (Bleek 1924: unnumbered page).

The condescension, paternalism, and triviality implicit in Dorothea's words clearly derived from her apparently ineluctable colonial perspective (Yates, Parkington, & Manhire 1990:21). Nevertheless, she, like Lloyd, had great affection for the /Xam people and certainly did not despise them. On the contrary, she devoted her entire life to curating, publishing, and studying the archive she inherited from her father and aunt. She also conducted her own San fieldwork, an arduous undertaking at that time.[14] When she died in 1948, she was drawing on the family archive to compile *A Bushman Dictionary*; it was published posthumously (Bleek 1956).

Despite Dorothea's diminishing view of San people and how it may have affected her handling of her father's and aunt's texts, it would be wrong, as was at one time the case, to assume that we cannot derive much useful information from the Bleek and Lloyd Archive: there is more in the texts than the Bleek family themselves realised.

Considering Myth

In setting about that task, I do not follow academic custom and begin by drawing distinctions between 'myth', 'legend', 'fable', 'folktale', and so forth (cf. Kirk 1974:18–21). I find that these usual 'universal' categories are not helpful in discussions of San narratives and lore.[15] They automatically impose Western expectations and, consequently, significances on the texts and so mask indigenous ones. The problem here is that, fundamentally, any process of classification is in itself a form of interpretation, not a neutral prelude to interpretation. Significantly, the /Xam used only one word, *kukummi* (sing. *kum*), to mean stories, news, talk, information, history, and what we call myths and folklore (Bleek 1956:106).

Nor do I attempt a general overview of 'San mythology': summaries of this kind are inevitably superficial and tendentious. They start from the summariser's expectations. Instead, and not underestimating my own expectations, I select a few southern San narratives for 'excavation' to show how each hangs together and affords insights into San thinking on particular topics. In doing so, I try to dig down to the connotations of the /Xam's own indigenous words. This sort of work confirms my view that a /Xam narrative cannot be said to have only one monolithic meaning that is conveyed by a didactic plot (for example, this is how the mountain came to be created). In any event, other San tales that I do not mention deal primarily with different issues from the ones I do discuss. My ignoring of them does not mean that I consider them unimportant. For example, some deal with relations between girls at puberty ('new maidens' in /Xam idiom) and supernatural entities and transformations; others deal with gender relations.[16] Another example is the ubiquitous, polysemic story about the dispute between the Moon and the Hare (Guenther 1999:126–45). Acknowledging this wide variety, I do not claim to have exhausted all the meanings that may be inherent in even the few texts on which I focus: they are rich beyond the endeavours of a single analyst.

We are sometimes told that, despite Bleek and Lloyd's meticulous work, the original indigenous meanings (however we may think of 'meaning') of these /Xam texts have been lost and that we can only speculate what they were. That view derives, in the first place, from a superficial evaluation of San (and other) ethnography as a category of evidence. Ethnographies are compiled in different ways, some more reliable than others; all are responses to whatever aim an interpreter has in mind. In particular, the dismissive view overlooks the numerous writers who, aware of its limitations, have carefully evaluated San ethnography in all its complexity.[17] One of the results of my own work has been that apparently simple texts such as the 'Song of the Broken String' are studded with far-reaching words and concepts that are unintelligible to, and therefore easily missed by, modern readers—as, I suspect, many were by the Bleek family themselves. These 'nuggets', as I call them, encapsulate meanings that bring San lore and myth to life. Specific narratives are seldom pan-San, but, as we shall see, nuggets frequently are.

Nuggets should not be confused with the cross-cultural narrative motifs (for example, Thompson & Roberts 1960) that, for instance, the folklorist Sigrid Schmidt (1989) used in her valuable catalogue of Khoesan folklore. Nor are nuggets equivalents of Claude Lévi-Strauss's 'mythemes' that, in his formulation, frequently comprise a subject and a predicate (Lévi-Strauss 1972:206–31). Rather, nuggets are single words denoting, for example, items of material culture that have rich associations, or parts of the natural environment with cryptic connotations. They may also be idiomatic turns of phrase that are opaque to outsiders, or ellipses that hearers would have been expected to complete from their own knowledge. Although diverse, nuggets are important because they invoke reticulations of fundamental beliefs and associations that may not be explicitly expressed in the text. As a narrative proceeds, they add up to a powerful, all-embracing cognitive and affective context. They provide a counterpoint to the manifest plot of a tale, enriching its harmonies and resonances. The manifest meaning, or 'lesson', of a narrative (if we assume one can be discerned) should be seen within this, for Westerners, allusive and often elusive context.

My use of the concept of nuggets explores, in part, the same territory as the notion of 'key symbols'. Although broader than key symbols, the notion of nuggets does imply a summarising or synthesising function. In Sherry Ortner's (1972:1344) words, they 'relate the respondent to the grounds of the system as a whole'. Respondents seldom analyse nuggets or key symbols, but they have absorbed their referents in the course of daily life.

Indeed, nuggets are part of the 'taken-for-granted' aspects of myth. Often indigenous narrators ignore the most important contexts and elements of a myth as being so obvious that they cannot imagine that their auditors do not think in terms of them. They themselves seldom, if ever, articulate them. In ancient Greece, for instance, writers and speakers rarely retold myths in detail. They more commonly merely referred to an incident or character in a myth on the assumption that their readers or hearers would know the full narrative (Kirk 1974).

Similarly with the San, we must constantly remember that in traditional circumstances the hearers were already familiar with the whole tale. They would mentally fill in 'missing' episodes or details as the narrator progressed. It was therefore not necessary for narrators to spell out every incident in the tales that they were performing. It was not even necessary that a tale be told through to its end: everyone knew how it ended (Barnard 2013:65). The taken-for-granted factor was high.

Within an encompassing intellectual universe like this, a small part, a nugget, can readily stand for a vast, unarticulated whole. Indeed, synecdoche is intrinsic to a San speaker's recounting and manipulation of narratives. In

Chapter 7 I argue that this principle applies, in modified form, to San image-making as well. An appreciation of nuggets soon destroys the illusion of simplicity in myth and art.

Although she does not use the word, Megan Biesele, who is fluent in the Ju/'hoan San language, realises the importance of nuggets as expressing unstated knowledge. Drawing on her decades of first-hand experience of the San, she contradicts Anne Solomon's (2009:33) defence of ignorance: 'Evaluation of the aptness of readings of the /Xam testimonies does not require extensive anthropological expertise'. The phrase that I have italicised in the following quotation is key to the approach that Biesele advocates: 'Picking out comparable "elements", however defined, becomes a challenging detective game involving much social knowledge and a wide grasp of the themes and plots in the body of narratives. Often a concrete detail, merely mentioned in passing, is enough to give *a social clue with enormous ramifications*' (Biesele 1996:145).

She points out that many San words 'are often extremely elliptical and could be very puzzling for some one unfamiliar with Zhū/twāsi [Ju/'hoan] language and culture' (Biesele 1975a:176). Similarly, Nicholas England (1968: 343–44), who studied San music, wrote of '"catch words", crammed full of meanings and surrounded by an extensive halo of implicit meanings'.

Such 'elements', or nuggets, may be easily missed, especially if they are part of a throw-away remark—'merely mentioned in passing'. Sometimes, of course, they are not mentioned at all, so obvious are they to indigenous listeners. Yet, severally and in combinations, they underpin and make *indigenous sense*, rather than some universal sense, of San mythical narratives. Some approaches to myths that do not follow this course lead to complex philosophical and psychological explications with a Western flavour that, although I do not necessarily denigrate all that sort of 'literary' work, seem to me to be far removed from anything that may have been in San minds.

Eric Csapo (2005:316) remarks that 'other people's myths become interesting and important when they are incorporated into one's own mythology'. He is right. Much writing on myth is the appropriation of other people's myths by the imposition of one's own values and interests. I believe that the myths of ancient Greece have suffered this fate. Identifying and explicating nuggets provides some defence against this form of appropriation. While narrative frameworks carry their own messages (the manifest meaning of a myth) and should not of course be ignored, nuggets activate the complex web of indigenous meanings in which the story resonates.

It is by 'excavating' nuggets that we can begin to hear something, if only a little, of what San people themselves 'heard' during a performance of a myth. A principle of communication applies here. Generally speaking, meaning is repeatedly created and recreated by the interaction of what is said with the content of listeners' minds. San minds were prepared, probably to varying degrees, to comprehend the nuggets of their myths. San narratives make sense only when seen within their own cosmology, social relations, and ritual —the dynamic, complex San *milieu*. We must therefore not take English recensions at face value; wherever possible, we must find out what connotations key indigenous words triggered in San minds.

These connotations are not all lost, though some undoubtedly are. They are frequently tucked away in the 12,000 pages of the Bleek and Lloyd Archive and in other nineteenth- and twentieth-century primary San sources to which I refer in subsequent chapters. Interpretations of San texts should be assessed not by the way they fit Western social or psychological theories of myth but, principally, by this sort of ethnographic explication. The evidence thus identified should not be simply assumed; it needs to be stated, wherever possible in the San's own words. Lévi-Strauss wrote: 'Against the theoretician, the observer should always have the last word, and against the observer, the native' (1978:7). He was perhaps skating over the difficulties involved in this process. Nevertheless, I try to give the San the 'last word'.

As I unpack, one after the other, the meanings of nuggets embedded in the myths that I have chosen, it becomes apparent that they are not isolated, one-off puzzles. Rather, they are parts of an overall pattern of thought. What may have seemed a fairly straightforward tale with a simple moral or 'lesson'—a naïve parable—becomes a thicket of interrelated meanings. Rather like words entered in a crossword puzzle, the meanings of nuggets overlap and run one into another. Thinking of his 'mythemes', Lévi-Strauss wrote: 'The true constituent units of a myth are not isolated relations but *bundles of such relations*, and it is only as bundles that these relations can be put to use and combined so as to produce meaning' (1972:211; his emphasis). The same is at least partly true of the interrelated nuggets that I identify in San myths. They are stand-alone entities or entity clusters, the constituents of which complement one another and thus provide extra facets of meaning that are not necessarily structuralist binary oppositions, such as life:death, male:female. Having been (reluctantly) drawn into such neat oppositions in his own work, Guenther concluded: 'I now have the uneasy sense that my analysis obscures as much as it explains' (1999:159).

Analysing Nuggets

Far from being the simple wandering hunters of the popular imagination, or as in Dorothea Bleek's view 'all his life a child', the San created a complex universe that interweaves acute observations of the natural and social worlds with elusive religious beliefs. This complexity is the foundation of their myths. When we look at the 'Song of the Broken String', we need to ask a number of questions that inevitably lead us to grapple with nuggets:

- Who are the 'people' mentioned in the song?
- What is the 'string' that they broke?
- Why did they break it?
- In what sense was the place thereafter not 'pleasant'?

I return to these four questions in Chapter 9. To answer them, we must first explore the thought-world and social circumstances of the San as they are manifest in their myths and lives. It is in this context that the song originated and in which it makes sense. As we proceed, we will see that the song is not a simple, straightforward lament. It is far more complex and ambiguous than it may at first seem.

CHAPTER TWO

Bringing Home the Honey

Have you not hunted and heard his cry?
(Qing in Orpen 1874:3)

Although Wilhelm Bleek and Lucy Lloyd's records are by far the most comprehensive, Diä!kwain and the other /Xam men and women were not the only nineteenth-century San to speak about their lives and beliefs. As preparation for the more complex /Xam texts that I discuss later, I base this chapter on two comparatively short tales given by Qing, one of those other San people. The record of this man's words makes for an interesting comparison with the facsimile pages in the Prologue: it is more like the unsatisfactory Western summaries of exotic myths with which many folklorists and social anthropologists have to deal than the Bleek family's verbatim phonetic transcriptions.

Nevertheless, it usefully introduces, in a compact way, some of the crucial themes and nuggets that we shall encounter in later chapters. As we saw in Chapter 1, the tales may appear baffling or trivial without some knowledge of the fundamentals of San society, religion, and ritual. I therefore give a brief sketch of southern San social and religious life and describe their central religious figure—the Mantis. He is the protagonist in the tales that I discuss in this and following chapters. Much in southern San life revolved around him. This chapter also opens up important questions about the mosaic of multiple San sources, scattered as they are through time and space. We need to ask: is it legitimate to cite twentieth-century Kalahari sources when explicating southern nineteenth-century records?

Myth and Meaning: San-Bushman Folklore in Global Context by J. D. Lewis-Williams, 47–74

A Glimpse

A curious accident rather than design led to the compiling of the source on which I now draw. It so happened that in 1873, the year that Diä!kwain went to live with the Bleek family in Cape Town, a colonial administrator by the name of Joseph Millerd Orpen (1828–1923) was called on to participate in a military expedition into the vast mountain ranges that lie some 200 km to the east of the plains where the /Xam lived (Figure 2.1).[1] These mountains are now known as the Maloti-Drakensberg or uKhahlamba. They are part of Lesotho and the southern and eastern border between Lesotho and South Africa (Colour Plate 2). In the nineteenth century, this was an area into which few colonists had ventured. Indeed, it is still regarded as somewhat remote. The purpose of Orpen's expedition was to capture Langalibalele, a Hlubi chief who had fled into the mountains to escape from the British colonial authorities in Natal. As it turned out, the expedition was a failure: Langalibalele was handed over to the British authorities by a Sotho chief far to

FIGURE 2.1 Joseph Millerd Orpen

the north of where Orpen was searching. In later years, Orpen (1964) seems to have avoided reference to the debacle; perhaps he regarded it as an embarrassment to his political career. He eventually became a respected member of the Cape Parliament and, finally, Minister of Lands and Agriculture in what was then Rhodesia, now Zimbabwe.

Orpen's first significant contact with the San probably coloured his attitude to the Maloti expedition. It occurred as early as 1855 when, as a young magistrate in the independent Orange Free State Republic, he was called on to raise a commando to resolve a dispute between some San and the Boer farmers on whose land they were living (Lewis-Williams 2003:99–103; Orpen 1964). Ill-advisedly listening to the colonists' urging, the inexperienced Orpen gave orders to attack the San. By the end of the engagement, ten San men and five women had been killed. In addition, three colonists were killed and five were wounded, including Orpen himself. Eventually, he learned that the whole affair had been engineered by a particularly bellicose Boer farmer who was not even the owner of the land. It was said that he had wanted to capture San children whom he could indenture to work for him, the form in which slavery survived in southern Africa. Orpen wrote that he was appalled by the tragedy and that he blamed himself for the loss of life.

Later, his interest in the San and their rock paintings was encouraged by the work that the Bleek family was doing in Cape Town and also by George Stow, a geologist who was making copies of San rock paintings and writing a book that was eventually titled *The Native Races of South Africa: A History of the Intrusion of the Hottentots and Bantu into the Hunting Grounds of the Bushmen, the Aborigines of the Country* (Stow 1905).[2] This long title sums up Stow's understanding of southern African history. His interest thus raised, Orpen gladly seized the opportunity that the Maloti expedition afforded to find out more about the remote mountain San and their beliefs and to discover whether they had anything in common with the Cape /Xam.

Orpen's source was Qing, a young San man who had 'escaped from the extermination of [his] remnant of a tribe in the Malutis [*sic*]'. He was the son of their chief and 'had never seen a white man but in fighting' (Orpen 1874:2). When Orpen met him he was employed by a Phuthi chief as a hunter, but the chief allowed him to guide Orpen through the uncharted territory. Unlike the San, who were traditionally hunters and gatherers, the BaPhuti are an agricultural people who live settled lives in villages. SePhuti was the language in which he spoke (Orpen 1874:3). He 'proved a diligent and useful guide, and became quite a favourite, he and his clever little mare, with which he dashed among the stones like a rabbit when his passion for hunting occasionally led him astray' (Orpen 1874:2).

As the military column laboriously journeyed through the mountains, Qing took Orpen to painted rock shelters and there, in the presence of the images, explained them to him. He also recounted a number of myths.[3] Orpen did not have the linguistic ability to record Qing's words in phonetic script. Instead, with the help of interpreters, he summarised the tales and made them 'consecutive' (Orpen 1874:3). As a result, it is today sometimes difficult to tell where one story ends and the next begins, although it has to be said that the San themselves probably did not think in terms of strictly demarcated stories. Despite these limitations, Orpen compiled an immensely valuable collection of San myths and folklore together with Qing's comments on rock paintings. It is important to remember that these tales and comments were contemporary with the making of the final images of the long San rock art tradition that had existed in the mountains for millennia.[4] Orpen wrote of only 'a glimpse into the mythology of the Maluti Bushmen', but he provided much more.

The Contact Period

For some centuries, the Maloti San had been in contact with Bantu-speaking agriculturalist neighbours with whom they interacted in various ways and to various degrees—the Sotho to the north of the Drakensberg and the Nguni to the south and east.[5] Comparable circumstances had arisen centuries before in the Kalahari and, in the 1990s, became the subject of what is now known as the Kalahari Debate.[6] Researchers asked to what extent the Kalahari San were dominated by Bantu-speaking agriculturalists, such as those now known as the Herero, Tswana, and Ovambo. The answer is not simple, but today, despite long interaction, the Kalahari San still consider themselves to be different from their agriculturalist neighbours. Even though the San may be to varying degrees economically dependent on them, they are not, as has been argued, merely a 'rural underclass' (Wilmsen 1989), itself a problematic term (Widlok 2004:223–24). In his review of the debate, Peter Mitchell concludes: 'The long-standing presence of food-producers in many parts of the Kalahari did not produce universal dependence, or result in the loss of distinctive hunter-gatherer languages, kinship systems, cosmologies, technologies and social systems emphasising sharing, reciprocity and egalitarian gender-relations' (2002:224).

I return to the Kalahari Debate in later sections of this chapter. For the present I note only that the principal Kalahari San religious beliefs are not derived from the agriculturalists. They do not, for instance, practise ancestor veneration. Then, too, Megan Biesele remarks: 'There is a general and very obvious separation between Bushman oral literature and that of the Bantu-speakers'(1993:31)—a significant point. The more recent Christian missionary influence, linked as it is to Western education and health-care, seems today to have been greater in some areas than the agriculturalists' influences (Guenther 1997, 1999:116–20).

To some extent the differences between the two sides in the Kalahari Debate concern, on the one hand, idealised concepts of what some researchers believe asymmetrical 'interaction' necessarily implies and, on the other, attention to empirical evidence. In the southeastern mountains, too, some writers have assumed that 'contact' and 'interaction' necessarily mean the virtual obliteration of foraging groups by 'more powerful' food producers. The empirical evidence suggests a more complex history, although extinction of the San way of life in the area from which Qing came was, by the end of the nineteenth century, indeed the final result.

Contact was characterised by, for instance, significant intermarriage between the two groups: usually a San woman married into an Nguni Bantu-speaking family. As a result, the Nguni, unlike the Kalahari Bantu-speaking agriculturalists, took over the distinctive clicks that characterise San languages (Louw 1975). In these circumstances, the San adopted numerous items of material culture from their neighbours (for example, Jolly 2006). In one of the myths that this chapter considers, the eagle's assegai (a spear with an iron point) is an instance. The Maloti San obtained broad-bladed iron points for assegais from the iron-smelting agriculturalists, and these weapons took their place alongside the San's traditional wooden spears. The /Xam, who lived to the west, had no immediately adjacent Bantu-speakers. They obtained iron by long-distance trade. In a /Xam creation myth, the first eland is also spoken of as having an assegai (Chapter 3; L.VIII.6.6533). Nevertheless, it seems that, in the west, the integrity of /Xam beliefs and rituals and their independence from the more easterly Bantu-speakers' beliefs is patent. But does 'borrowing' in the east imply that the San of the southern mountains where Qing lived went further and adopted the farmers' religious beliefs and myths to any marked degree?

Some light is shed on this question by one of the most illuminating, although brief, accounts of interaction between the eastern San and Bantu-speakers. Walter Stanford (1910) recorded it in 1884. In the final decades of the nineteenth century an Nguni man named Silayi went to live with a predominantly San nomadic group who, together with people of mixed origins, occupied Drakensberg rock shelters (for example, Blundell 2004; Challis 2012). The San leaders of the group received him cordially and presented him with bows and arrows. Did he know how to use them, or were they merely symbolic? I suspect they were principally emblems of membership of a predominately San band. Although they were living in an itinerant, culturally mixed group that Sam Challis (2012, 2014) has described as 'creolised', Silayi reported that the San had not adopted agriculturalists' characteristic customs, such as male initiation ceremonies.

According to Silayi and other sources (for instance, Callaway 1969:10; Report 1883:409), southern San ritual specialists used their traditional rites to make rain for the agriculturalists. When Stanford (1910:438–39) asked Silayi if the San had 'witch doctors' (a word associated with Bantu-speakers), he denied it and said that they had 'rain doctors'. Silayi was thus explicitly distinguishing between traditional San ritual specialists and Bantu-speaking specialists. In this context, 'borrowing' moved from the San to the southern Nguni Bantu-speakers to the extent that the agriculturalists took over the San word *!gi:xa* to denote their own ritual specialists. They also selectively adopted, in modified forms, a number of other San religious elements, including *thwasa,* a 'trance-like' state of dreams, visions, and psychic experiences that became part of a Cape Nguni diviner's 'consecration'. At the same time, the San trickster /Kaggen became the Nguni witch familiar *Thikoloshe,* a figure who has spread far and wide.[7] In addition, the San with whom Silayi lived made rock paintings, unlike the agriculturalists. He commented: 'They could paint very well' (Stanford 1910:439). The San painting tradition had not completely died out as a result of contact with other people. In some instances, though, individuals and small groups adjusted to the new conditions, and this development led to other, highly ephemeral, types of painting.[8] We should expect to find both continuity and change in contact situations like this.[9] The nineteenth-century mixed groups varied: some comprised people who held different beliefs; others included creolised people of mixed beliefs. The degree and nature of 'borrowing' therefore probably varied from individual to individual rather than from one homogeneous group to another. The groups were, moreover, comparatively short-lived and fluid, perhaps lasting for only a decade or two. By contrast, one lifetime could cover the period from, say, 1850 to the time of Qing's meeting with Orpen and then well into the twentieth century. Careful discriminations must therefore be made, and blanket judgements should be avoided: as I again point out in Chapter 8, the southern San, as a group, were not entirely sucked into Bantu-speaking culture.

Orpen and Qing

Orpen obtained Qing's tales in informal circumstances 'when happy and at ease smoking over camp-fires' (Orpen 1874:2), not in authentic San performances in which the narrators would have dramatised, expanded, and varied the stories to suit particular occasions (cf. Biesele 1993:27–31). There are therefore no definitive versions of the Maloti narratives—or indeed of any San narratives, the /Xam tales included. Not expecting the different variants that he was hearing, Orpen was puzzled. Writing of what he called Qing's 'fragmentary stories' he said: 'They either varied a little, or I failed to understand him accurately when speaking through different translators.' He also wondered if Qing's youthfulness meant that he had not 'learnt them well' (Orpen 1874:3). Today it seems more likely that Orpen did not understand the fluidity of San mythology: ironically, 'fragmentary' sums up a characteristic of episodic San myths. Indeed, new rock art evidence suggests that Qing was more than familiar with the beliefs of his own people (Challis, Hollmann, & McGranaghan 2013; Lewis-Williams & Challis 2011:115–18).

Immediately after the Maloti expedition, Orpen wrote the article that includes his collection of myths and his copies of rock paintings. He sent it to the editor of *The Cape Monthly Magazine* in Cape Town, and the editor passed it on to Wilhelm Bleek for comment. It was published in 1874 along with an addendum by Bleek (Figure 2.2).[10] The manuscript is now preserved in the South African Library in Cape Town.

Reading the manuscript before its publication, Bleek found that there were differences between the myths of the two regions: 'There is not one of [Orpen's] myths which is exactly identical with any one of ours' (Bleek 1874:11). Nevertheless—and this is the key point—Bleek detected intriguing parallels, some of which I identify in this chapter. He concluded that, despite the differences, 'the general character of the myths recorded by Mr. Orpen is mainly the same as that of those collected by us' (Bleek 1874:11). He also found that 'different circles of myths' seem to surround 'the same central figure' in the mythologies of the two San groups—the Mantis. Although the name of this being was identical in both regions, Bleek suspected that the

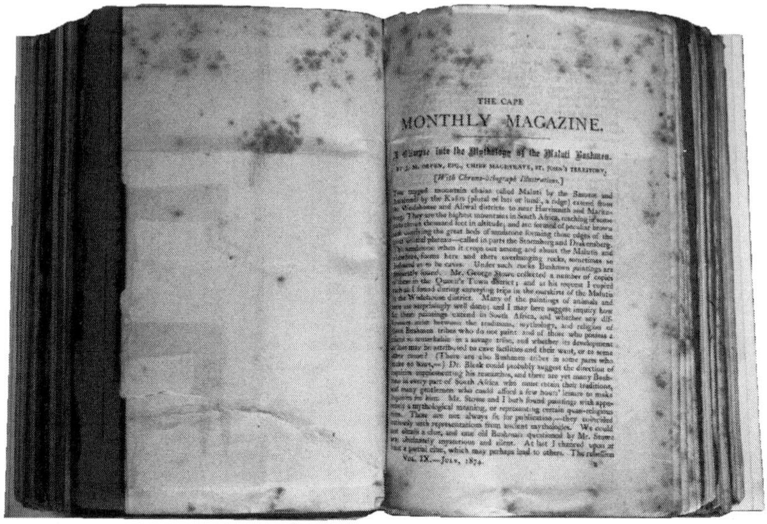

FIGURE 2.2 Joseph Orpen's 1874 article appeared in the Cape Monthly Magazine.

Maloti figure was rather more beneficent that that of the /Xam (Bleek 1874:11). Had he lived beyond 1875, Bleek may well have revised this judgement as more /Xam narratives became known (Lewis-Williams 1981:122).

Two Maloti San Myths

What we have in Orpen's record are basic, pared-down versions of Qing's narrations. To understand their embracing thought-context we have to go beyond the narratives themselves and look for nuggets embedded in them. Many were doubtless lost in Orpen's summarising, but some did survive. It is in them that we find indications of 'the general character' of San myths of which Wilhelm Bleek wrote. This form of literary excavation leads us, point by point, toward an understanding of the ways in which San myth, ritual, and belief are interrelated.

The two complementary narratives that I have selected are separated in Orpen's article by a seemingly disparate tale. Presumably, he inserted it to give the impression that some time elapsed between the events of the two adjacent narratives (Orpen 1874:8–9). There are no breaks in Orpen's publication or in his manuscript. For the sake of completeness, I provide the intervening tale in a smaller typeface; I do not discuss it. By adding paragraph breaks and adjusting punctuation to facilitate easy reading, I have created a lightly edited version of Orpen's rather crowded published format. I have not eliminated or added words or phrases. Orpen's complete article is now available to researchers on line (McGranaghan, Challis, & Lewis-Williams 2013). It is a mine of intriguing information.

The protagonist in both tales is the 'central figure' of whom Bleek wrote. In his orthography, the name is /Kaggen, the solidus or vertical line (dental click) taking the place of the C in Cagn in the Bantu-language orthography that Orpen used. The final syllable in /Kaggen is very short, hardly pronounced at all. As Bleek (1874:11) pointed out, the pronunciation of /Kaggen and Cagn is the same. This is the name that is usually, and now famously, translated as 'the Mantis'. Cogaz, who is referred to as a 'great chief', is Cagn's elder son. Qüuisi is not mentioned elsewhere in the collection. The titles are mine.

CAGN & THE HONEY: I

Cagn found an eagle getting honey from a precipice and said, 'My friend, give me some too.'

And it said, 'Wait a bit'. And it took a comb and put it down, and went back and took more, and told Cagn to take the rest.

And he climbed up and licked only what remained on the rock. And when he tried to come down he found he could not. Presently, he thought of his charms, and took some from his belt, and caused them to go to Cogaz to ask advice. And Cogaz sent word back by means of the charms that he was to make water to run down the rock, and he would find himself able to come down.

And he did so, and when he got down, he descended into the ground and came up again. And he did this three times, and the third time he came up near the eagle, in the form of a large bull eland.

And the eagle said, 'What a big eland' and went to kill it. And it threw an assegai, which passed it on the right side, and then another, which missed it, to the left, and a third, which passed between its legs.

And the eagle trampled on it, and immediately hail fell and stunned the eagle, and Cagn killed it, and took some of the honey home to Cogaz, and told him he had killed the eagle, which had acted treacherously to him. And Cogaz said, 'You will get harm someday by these fightings'.

CAGN & CGORIÖINSI

And Cagn found a woman named Cgoriöinsi, who eats men. And she had made a big fire, and was dancing round it, and she used to seize men and throw them into the fire. And Cagn began to roast roots at the fire.

And at last she came and pitched him in, but he slipped through at the other side, and went on roasting and eating his roots. And she pitched him in again and again, and he said only, 'Wait a bit till I have finished my roots, and I'll show you what I am'.

And when he had done, he threw her in the fire as a punishment for killing people.

CAGN & THE HONEY: II

Then Cagn went back to the mountain, where he had left some of the honey he took from the eagle. And he left his sticks there and went down to the river. And there was a person in the river named Qüuisi, who had been standing there a long time, something having caught him by the foot and held him there since the winter. And he called to Cagn to come and help him.

And Cagn went to help him and put his hand down into the water to loosen his leg. And the thing let go the man's leg and seized Cagn's arm. And the man ran stumbling out of the water, for his leg was stiffened by his being so long held fast, and he called out, 'Now you will be held there till the winter'.

And he went to the honey and threw Cagn's sticks away.

And Cagn began to bethink him of his charms, and he sent to ask Cogaz for advice through his charms. And Cogaz sent word and told him to let down a piece of his garment into the water alongside his hand. And he did so, and the thing let go his hand and seized his garment. And he cut off the end of his garment and ran and collected his sticks, and pursued the man and killed him, and took the honey to Cogaz.

Narrative parallels between the two flanking tales are clear:

- The central dispute concerns possession of honey.
- Cagn gets into difficulties when he tries to obtain honey.
- He seeks Cogaz's advice by means of his 'charms'.
- He gains access to the honey and takes it home.
- He kills the 'spoiler'.

To appreciate that these two tales are more than simply amusing stories about the picaresque Mantis we need to know something about San values and personal relationships—neither are the same as those that researchers know if they are from a non-San background. For instance, at the beginning of the first tale, Cagn refers to the eagle as 'my friend' when he asks for some of the honey. At first glance this form of address may seem merely amicable or, perhaps, ingratiating. But the salutation is a nugget. It invokes sharing relationships that underlie San society and many myths. All San tales should be read in terms of these relationships. To understand them, we need to expand our southern nineteenth-century sources by turning to the San who still live in parts of the Kalahari Desert in Namibia and Botswana and were the subject of the Kalahari Debate.

Widely Separated San Groups

Since the 1950s, when the Marshall family and a group of researchers from Harvard University began to study them, the Kalahari San have become one of the best known hunter-gatherer peoples in the world (Gordon 1986). They include, for example, the !Kung, or Ju/'hoansi, the !Kõ, the Nharo, and the G/wi (Figure 1.1). Today, most of these communities are to some extent acculturated, as the Kalahari Debate and other research has shown, but numerous groups still preserve old beliefs and practise traditional rituals. Unfortunately, finding the 'last pure Bushman band' has become a misleading journalistic trope (Gordon 1992).

Elsewhere I have discussed in detail the relationship between the Kalahari and the southern San sources, separated as they are by a century and a quarter and a thousand or more kilometres, as well as by linguistic and ecological divisions.[11] Having found detailed parallels, I have revised my earlier view that researchers should not draw on the northern ethnography when they are trying to understand nineteenth-century southern beliefs and rock art.[12] I now conclude that the recent Kalahari material can be used to explicate the comparatively limited, although none the less considerable, nineteenth-century southern San records *in those conceptual areas where fundamental parallels can be demonstrated.*[13] The nineteenth-century and the Kalahari San ethnographies, although by no means identical, are, in some respects, complementary. The differences between them do not cancel out the parallels.

I am not alone in taking this view. Indeed, anthropologists have documented the essential unity of fundamental San religious beliefs and practices, the domain with which we are now principally concerned.[14] Mathias Guenther, for instance, has drawn attention to this point: 'The special status some might want to ascribe to the myth and lore of the /Xam Bushmen—for instance, Schapera (1930:398), who deemed the southern Bushmen to "stand apart from the rest"—is seen to dissolve and the fact that Bushman expressive and religious culture does indeed constitute one unit (Guenther 1989:33–36) becomes the more apparent' (1996:98).[15]

As Guenther and other writers imply, there were not only social but also 'expressive and religious' parallels between the major San linguistic groups. Some of these parallels are fundamental to an understanding of Qing's tales and indeed all San folklore. I deal with them in detail shortly. More of them will become apparent as we proceed through subsequent chapters.

The Social Setting of Qing's Tales

The nineteenth- and the twentieth-century ethnographies both show that the San lived in small, mobile bands of fluid membership, each band comprising only a few nuclear families. These bands did not have 'chiefs', and decisions were taken consensually.[16] The small bands, which moved seasonally, were parts of an extensive and fluid 'band nexus' that was held together by kinship (consanguineal and affinal), fictive kinship (based on duplicated names), and personal exchange partnerships that, with delayed reciprocity, could last for many years.[17] The popular notion of isolated hunting bands fiercely defending a demarcated territory has no foundation. Toward the end of the nineteenth century prominent leaders seem to have emerged in various San groups, but especially in the region of the Maloti-Drakensberg, Qing's homeland.[18] The San should not be assumed to have no history.

All the relationships I have mentioned constitute a strong moral framework that extends across bands and ensures equitable distribution of scarce resources (Keenan 1981). Sharing is fundamental to San morality. Even in the nineteenth century, the southern San were noted for their ethic of sharing. As an early writer observed: 'They would always share their food with others, even when it was scarce' (Ellenberger & Macgregor 1912:7). More recently, Richard Lee, writing of the Kalahari Ju/'hoansi, describes sharing as 'one of the core values of the ideological system' (Lee 1979:156). Of, specifically, meat sharing, Lorna Marshall writes: 'The idea of sharing is deeply implanted and very successfully imposes its restraints' (Marshall 1976:303). Having studied a Kalahari San group near Kutse in Botswana, Susan Kent (1993:506) sums up the situation: 'sharing is the basic adhesive that holds the society together'.

It is clear that this concept of sharing would have been in the minds of San people who listened to Qing's two tales and would have contextualised what was happening with the honey. Certainly, 'sharing' had much richer connotations for the San than the word has for English speakers. The Mantis's salutation 'my friend' invokes those specifically San relationships.

Apart from using the phrase 'my friend', Qing did not clarify the specific relationship that Cagn was claiming when he asked the eagle to share its honey, but, in asking rather than simply taking, he was following approved San procedure. Although permission is seldom, if ever, refused, one should always ask before taking another band's or person's resources, be they water, plant foods, or game. The treacherous eagle, however, broke with accepted tradition, selfishly took most of the honey and left 'what remained on the rock' for Cagn to lick up—a niggardly portion, all the desirable combs having been taken. Already we can see that Qing's two tales about honey, far from being merely amusing stories, concern a central element in San ideology: sharing. But there is much more to the tales than a simple moral precept. We have not yet examined the tales' other nuggets.

The Religious Setting of Qing's Tales

The 'general character' that Bleek detected between the myths of the two nineteenth-century southern San groups extends much farther to the north. Alan Barnard (2007:96), for instance, found that 'religion is far more uniform throughout Bushman and even Khoisan southern Africa than are material aspects of culture and society'.[19] As we have seen, Guenther concurs: 'Bushman expressive and religious culture does indeed constitute one unit' (Guenther 1996:98; see also Guenther 1989:33–36). Although I continue to use the word, we should remember that 'religion' is not a neat given. It is a construct put together by scholars, a construct that is more easily seen as a bounded entity in some cultures than in others, the San being one of the latter.

This is an important point that some writers ignore or downplay. For reasons that seem to flow from their essentially Western viewpoint they discount the prominent role of San religion and, as I go on to show, specifically the San's principal religious experience: the altered states of consciousness that grade in intensity and that routinely occur in the frequently performed trance dance and other circumstances, such as dreaming and 'special curings' (for example, Guenther 2014). Perhaps uncomfortable with this form of mysticism and, one must note, the ways in which its experiences and imagery infiltrate other components of San life, these writers ignore the reports of ethnographers who have lived with the San. They consequently fail to see into the heart and functioning of San society. To ignore or marginalise these experiences in their variety and power is an assault on the integrity of the fullness of San life. As this book proceeds, the ubiquity and centrality of San trance experience will become evident: San ethnography is abundantly clear on this point.

Indeed, the healing, or trance, dance is, as Guenther puts it, 'the central ritual of the Bushman religion and its defining institution' (Guenther 1999: 181). It has been described and discussed many times.[20] Very briefly, the women sit around a central fire and clap and sing 'medicine songs' believed to contain supernatural potency. The men, half of whom at any given time may be ritual specialists or healers, dance in a circle around the women, now clockwise, now anticlockwise. Their dancing produces a marked circular rut in the sand. About a third of the women try to become healers, but only about 10 percent succeed (Katz 1982:160). Ju/'hoansi believe that during childbearing years women should not engage in altered state healing. Thus the 10 percent figure reflects not a struggle to succeed but fewer years available to try (Biesele, pers. comm.). Without recourse to hallucinogens, some of the ritual specialists enter a frenzied trance. Those who have learned to control their level of trance move around laying hands on all present to remove sickness, perceived or unperceived, from their bodies. In trance (as in dreams), they go on out-of-body travel and protect their people from malevolent spirits.

Although contact with a supernatural realm is central, the trance dance should not be considered a monolithic, unchanging ritual. Richard Lee (1993:119–21) describes the women's Drum Dance: unknown in the 1950s, it become popular in more recent times in some parts of the Kalahari. It was originally seen as training for aspirant female ritual specialists, though it came to rival the men's trance dance in popularity by the 1960s. One of the originators of the Drum Dance, a woman named /Twa, was still alive at that time. Similarly, potent new trance dance songs seem to originate sometimes with known individuals and then become widespread, as did the Giraffe Song that came in a dream to Beh, a Ju/'hoan woman, although there are different versions of her epiphany (Biesele 1993:70). Southern rock paintings also suggest that there were various trance dance choreographies (Lewis-Williams & Pearce 2004a:97–100; 2012). In San life there is a constant interplay between individuals and overall beliefs.

Contrary to an unfortunate misapprehension (Bahn 2010:101; Skotnes 1996b:238; Solomon 2007:157; Wessels 2010:277), the southern San did indeed practise trance dancing, though, given the disintegrating small groups to which they were reduced in the mid- and late nineteenth century, they probably performed them less frequently than the northern people still do (Lewis-Williams & Pearce 2012). Richard Lee (1993:116) found that the frequency of dances among the Ju/'hoansi depended on 'the season, the size of the camp and other factors' but mostly involved large numbers of people. A large dance may comprise fifty to eighty people (Katz 1982:38). Also among the Ju/ 'hoansi, the Marshalls observed 'one intense and ardent little dance' performed by eight people (Marshall 1999:64), but that was unusual.

In the second half of the nineteenth century George Stow found unmistakeable evidence for dancing in every camp and large rock shelter that he visited: 'The universality of this custom was shown by the fact that, in the early days, in the centre of every village or kraal, or near every rock-shelter, and in every great cave, there was a large circular ring where either the ground or grass was beaten flat and bare, from the frequent and constant repetition of their terpsichorean exercises' (Stow 1905:111).

Writing about the trance dance, Megan Biesele notes that 'the central religious experiences of Ju/'hoan life are consciously and, as a matter of course, approached through the avenue of trance' (Biesele 1993:70). Similarly, Guenther concludes: 'The fact that trance dances are described by all writers who have visited the Bushmen, even nineteenth-century ones, further attests to the ubiquity and antiquity of this key Bushman ritual' (Guenther

1999:181). He adds: 'In the fashion of shamans all over the world, the [San] trance dancer, by means of altered states, enters the spirit world and obtains from it the wherewithal to restore the health of sick fellow humans' (Guenther 1999:186; see also Guenther 1989). Although 'shamanism' covers a range of varieties worldwide, in anthropological discourse shamans are generally thought to be people who, by their own volition, enter an altered state of consciousness to make contact with a nonordinary world on behalf of their community. It has long been recognised that within that broad definition there are many 'shamanisms' (Atkinson 1992).[21] It is therefore significant there can be no doubt that the trance dance is central to San religious experience; it is not a religious sideshow. Indeed, Marshall (1969:373) wrote that, among the Ju/'hoansi, 'trance is a normal, habitual, common experience'.

Important for our understanding of the myths that I have selected, Biesele found that the metaphors of the trance dance, what I have termed nuggets, permeate Ju/'hoan folklore (Biesele 1993:83ff). It is here that we begin to find the elusive link between San myth, ritual, and religious experience. I emphasise the point: this underlying unity has been established empirically and, notwithstanding regional emphases, should not be ignored when researchers discuss San myth, ritual, and cosmology.[22] Nor should it come as a surprise that trance dances, or elements thereof, are frequently depicted in San rock art (Chapter 7). This does not mean that every image is a recorded vision: the situation is far more complex.

The Mantis

The religious dimensions of the tales that I discuss in this and subsequent chapters are especially signalled by the role that the Mantis plays. Indeed, all the tales I have selected for discussion in this book feature the Mantis. As I have mentioned, his name may be given as Cagn or /Kaggen. Though the orthographies are interchangeable, I reserve a particular spelling for its original published context. An understanding of the Mantis is fundamental to 'decoding' the myths that I have selected.

Early Western writers, some of whom thought that the San had no religion at all (Tindall 1856:26), tended to translate this figure's San name as 'Devil', 'Satan', or 'evil deity'. Indeed, reports of San belief in an evil being date from the eighteenth century (Sparrman 1789:1:148; Thunberg 1796:2:163) and continued right up to the beginning of the twentieth century (Curlé 1913:117). Wilhelm Bleek (1875:9) thought that the Boer farmers originated the satanic 'translation' of the name. The use of such words appears, for example, in a little-known 'Contribution from a Bushman'. Published in the *Orange Free State Monthly Magazine* in 1877, it gives us some idea of the dilemma in which the San of that violent time found themselves. According to Stow (1905:103, 134) Kina-ha spent some time 'under missionary instruction' at the Bethulie Mission Station in present-day Free State Province, South Africa. His remarks should therefore be read with this influence in mind. Kina-ha (1877:83) distinguished two personages, a benign spirit, T'koo, and a 'wicked spirit', T'ang. Early writers often used an apostrophe to indicate a click of any kind. The T'koo of Kina-ha's report is probably not San but a form of the IsiXhosa word Tixo, God. T'ang, however is clearly a transcription of the more generally reported San name /Kaggen, inappropriately rendered by the various English demonic terms. Kina-ha, living in a missionary context, was probably trying to reach an accommodation and please listeners who did not understand (or care about) the ambivalence of the San trickster.

The /Xam word */kaggen* may indeed be translated as 'praying mantis', but the being seems to have been only a little more thought of as an insect than Christians think of Christ as a lamb. There has been some debate as to whether 'trickster' is an appropriate term to describe him.[23] Without necessarily invoking that problematic category, we can be specific and say that the Mantis was ambivalent and paradoxical. He could be good and mischievous, wise and foolish; he could help people, but he also tricked them. Although people prayed to him to deliver an antelope into their hands, he also sometimes resorted to various ruses to cause an animal to escape from the hunters (Bleek 1932:233–40). His contradictory traits may appear in a single story. He is, moreover, frequently boastful, a deplorable trait in San eyes; but it was he who benevolently created all things: 'the sun, moon, stars, wind, mountains, and animals' (Orpen 1874:3; Lewis-Williams 1981:117–26).

All this may be confusing, but we should not reject what we know about /Xam beliefs about /Kaggen as a hopeless muddle created by late nineteenth-century informants' poorly retained reminiscences of an earlier time when beliefs were more consistent. On the contrary, in his person /Kaggen held together ambiguities and contradictions in a way that the San accepted as true and indicative of the essence of life (Guenther 1999:101–09). This is not to say that the Mantis 'symbolised' some abstract concept that the San consciously entertained. Life simply *is* unpredictable, ambiguous, and elusive—like the Mantis himself. In any event, we must remember that societies place different values on deviant behaviour. As Thomas Beidelman (1980:35) has pointed out, in some societies 'figures of disorder and deviance stand at the centre of the belief system, being gods activating key cosmological events'. We must beware of ethnocentric evaluations of deviance and contradiction.

In the numerous myths in which he appears, the Mantis behaves like an ordinary San person: he has a man's body, hunting equipment, and clothing. Nevertheless, he can transform himself into various creatures, including a louse, a snake, a hare, a hartebeest, an eland, and 'a little green thing' that flies, that is, presumably, a praying mantis.[24] When speaking of the Mantis in 1870s Cape Town, the /Xam man /Han≠kass'o (Chapter 3) was reminded of masked children celebrating the 5th of November: 'The Mantis imitates what people do also, when they want us who do not know Guy Fox [*sic*] to be afraid. They change their faces, for they want us who do not know to think it is not a person. The Mantis also . . . cheats them that we may not know that it is he' (L.VIII.17.5434).

This ability to change form is not shared by any other southern San mythical being, though in the Maloti tales Qwanciqutshaa, in some ways Cagn's alter ego, can change into a snake (Lewis-Williams 2013b; Orpen 1874:7). That the Mantis was so protean a figure may be one of the reasons why there are no convincing rock paintings of praying mantises: although we cannot be sure, he was perhaps too elusive for depiction. However, his absence from the imagery may simply be the same as the absence of other mythical characters: the art is not a straightforward depiction of myths and mythical beings.

As will become apparent, there is a further aspect of Cagn that is central to an understanding of Qing's tales, as well as the /Xam Mantis myths that I consider in later chapters: he was the original *!gi:xa*. This /Xam word has been translated as 'healer', 'medicine man', and 'doctor'. In 1874, Wilhelm Bleek struggled with 'enchanter' and 'wizard' (B.XXVI.2463') and, in conversation with Bleek, Diä!kwain suggested a Dutch word that Bleek recorded as *perfit*. An adjacent note suggests that this is actually a form of the later Afrikaans word *profeet*, 'prophet'. Eventually, Bleek and Lloyd settled on the English word 'sorcerer', but it too is now generally considered unsatisfactory.[25] The two syllables of the /Xam word hold the key to its fundamental meaning. The first syllable, *!gi:*, means supernatural potency. The second, *-xa*, means 'full of'. So, literally, *!gi:xa* (pl. *!gi:ten*) means 'one who is full of supernatural potency'. There is no San word for these ritual specialists that is common to all languages, although the Kalahari Ju/'hoansi do use *!gi:xa* (*g!aeha* in Ju/'hoan orthography) to mean a powerful 'healer' (Katz, Biesele,

& St. Denis 1997:58, 110). This is an interesting parallel between two San languages that are mainly lexically distinct. The more usual Ju/'hoan word is *n/om k"xau* (pl. *n/om k"xausi*), which means 'owner of potency'. The words *!gi:* and *n/om* both mean supernatural potency. Lorna Marshall compares potency to electricity. Some writers refer to it as 'medicine'. Though generally invisible, it can be seen and 'picked up' by people in trance. Katz (1982:94) describes it as the 'primary force' in the 'universe of experience'. Certainly, it is of major significance in San thought, myth, ritual, and art.

In itself, potency is neither good nor bad. Its concentration, context, and development are what matter: an intense concentration is dangerous. San ritual specialists, the *!gi:ten* or *n/om k"xausi,* possess and harness potency to enter the spirit realm and perform various tasks, which include driving away malevolent spirits, healing, visiting other camps by extracorporeal travel, and, among the /Xam, rain-control. They do all this in a trance dance, a dream, or in the more restricted circumstances of 'special curings', in which a sick person is treated without a full trance dance. The words *!gi:xa* and *n/om k"xau* are nowadays often translated as 'shaman', the word I and others use without implying identity with the 'classic' shamanism of Siberia.[26] In addition to being able to transform himself into a number of creatures, the Mantis can fly through the air and dive into waterholes, both shamanic feats. As we shall see in Chapter 4, to accomplish these miracles he 'trembles' (*!khauken*). This is what San shamans do when they enter trance. He also foretells the future by dreaming—again, as shamans do. The principal task of San shamans is, however, healing. In a clear act of San shamanic healing in a /Xam tale, the Mantis calms the Blue Crane by rubbing her with sweat from his armpit: 'He rubbed himself under the arms with his hands, and (then) rubbed the Blue Crane's face, he made the Blue Crane smell his odour' (L.VIII.32.8809). The /Xam had a special word for what was for them an important, significant action: *//Λ hóobaken,* 'to anoint, rub with perspiration' (Bleek 1956: 628). Dorothea Bleek, not knowing that sweat from armpits was significant in a San trance dance because it was believed to be intensely imbued with potency, omitted the relevant sentence from her published version (Bleek 1924:27; Lewis-Williams 1997:209). A nugget was thus overlooked.

Like a 'Lord of the Animals' in other cultures, the Mantis was believed to possess all large antelope and thus their supernatural potency: he controlled them as /Xam 'shamans of the game' (*øpwaiten-ka !gi:ten*) did when, by means of their supernatural abilities, they guided the movements of antelope herds in the direction of hidden hunters. When Orpen asked Qing where Cagn was, he replied: 'Where he is, eland are in droves like cattle' (Orpen 1874:3). It was Cagn's cry that alerted these 'droves' to the danger of approaching hunters. As later chapters show, herds of eland were reservoirs of potency, as well as potential food.

In the first of Qing's myths, Cagn's shamanic persona clearly comes to the fore. He 'descends into the ground' and comes up again three times—a San shamanic form of supernatural travel. In an account of such an experience a Ju/'hoan shaman told Biesele: 'My protector [a giraffe] told me that I would enter the earth. That I would travel far through the earth and then emerge at another place' (Biesele 1993:72; Katz 1982:113–14). By this and other means, Cagn is enabled to find out what is happening at a distant camp, an abiding San concern (Chapter 9). Underground travel is a nugget that points to his special power and accomplishments. In both tales, he achieves this by means of his charms, a point on which I comment later.

According to the /Xam, /Kaggen had a second name that was used by his wife: //Kanndoro, translated by Lloyd as 'tinder box owner'.[27] Although her choice of word is understandable, this is not an apt translation, because no container is denoted by the word. The first syllable, *//kan,* means 'to own' (Bleek 1956:557), and *doro* means the 'fire stick' that the San expertly twirl to light a fire.[28] At first, the name //Kanndoro may seem to recall Prometheus, the bringer of fire, and be part of a creation cycle such as is found in numerous societies. While that meaning may well be present, fire has more specific significances for the San. In eastern Kalahari San languages, *doro* means both 'mantis' and 'nose bleed' (Traill, pers. comm.). This is another link to trance, because southern San shamans bled from the nose, an occurrence often depicted in the rock paintings.[29] The Ju/'hoansi also speak of shamans' nasal blood (Marshall 1969:374). By a chain of associations, the name //Kanndoro thus relates the Mantis once more to the trance dance.

Absent from the nineteenth-century southern trickster material are the ribaldry and sexual allusions that characterise the more recent Kalahari records. The Bleek family may well have forbidden or discouraged such elements. Otherwise comparable beliefs about the trickster and shamanism still exist among the Kalahari Ju/'hoansi. One of their names for their trickster figure is 'G!ara'. In a Ju/'hoan myth, G!ara enters trance for the first time and thus creates the foundation for San spirituality and the social renovation with which shamans concern themselves. After some failed attempts, he puts eland fat (another point to which we shall return) and glowing coals into a tortoise shell, such as Ju/'hoan shamans still carry during a trance dance. The nineteenth-century /Xam used tortoise shells in a similar way (Bleek 1874:12). G!ara then 'sniffed the smoke, and this time he began to trance' (Biesele 1993:105). Like the southern San Mantis, the Ju/'hoan trickster was the first trancer. When the Bleek family had an opportunity to interview some young Ju/'hoan boys, they found that the trickster (/Xue, in the boys' version) could make rain (Bleek 1935b:269), as do some /Xam shamans. The essential unity of San belief across linguistic and geographical divides on which anthropologists have commented is again evident.

Another link is that the Ju/'hoansi have a trance dance song called !Garama ¯ku da. It means 'God-little is fire' (England 1968:453). The central fire at trance dances around which the women sit and the men dance in a circle is an important element because it heats the potency in the dancers, eventually causing it to 'boil' (Marshall 1962:250;1966:272; Biesele 1993:84, 97):

> In the immense darkness of space, the fire reaches out in a circle. It holds the singers and the dancers together and helps transform them for their healing task. . . . There is num [potency] in the fire, and the !Kung [Ju/'hoansi] work with the fire. . . . Healers use the same word, daa, to describe both the central dance fire and the fire within their own bodies. (Katz 1982:120, 121)

On especially memorable occasions, the central dance fire seems to overwhelm a shaman, and he sees potent animals in the fire: 'God is the one who shows you these animals in the fire. When you see the animal in the fire, it changes you. You become that animal' (Keeney 2003:85). Here we see that transformation into an animal is a feature of San trance experiences and myths. The Mantis has much in common with shamans.

A network of associations spreads out from the Mantis to fire, to the trance dance, and to transformation into an animal. The richness of Qing's apparently simple tales is becoming apparent. But it introduces problems. Some modern readers of San texts find it hard to picture a myth set outside of a strictly moral universe (as they understand morality). They cannot take a myth seriously if it does not have a clear 'lesson' or 'message'. In the myths and fables of literate societies there is more regularity, more distinguishing between good and evil, as they are conceived in, for example, the Judeo-Christian tradition. In that tradition order and morality are desired norms. Although appropriate behaviour, such as sharing, is important in the San universe, caprice also plays a prominent role. In place of a God who is primarily interested in righteousness the /Xam had an ambivalent trickster. His mere presence lifts a tale from what some would call a folktale to a tale with far greater significance. Unlike Judeo-Christian mythology and teaching that tends to paper over, or simplify, the realities of life, San religion grapples with ambivalence.

Honey

The Mantis's dispute with the eagle over, specifically, honey is part of the religious setting of Qing's narrative. Honey, of course, plays an important role in many mythologies and religions. In ancient Greek myth, in an intriguing parallel with the San narrative that I discuss in the next chapter, Melissa feeds the infant Zeus honey and so contrives his escape from his murderous father Cronus. Then, in the Bible we read of a land flowing with milk and honey, and John the Baptist survived in the wilderness on honey and locusts (as we shall see, with locusts we have another curious parallel with San life and belief; McGranaghan 2012/4).

These and other contexts can all too easily lead us to impute to the San Western thoughts about honey. This sort of dilemma is at the root of understanding—or misunderstanding—many myths, those that we have inherited from ancient classical times and those that we encounter in other cultures, as well as our own. Before we start to unravel the 'lessons' or 'morals' supposedly conveyed by myths, we must first unpack the nuggets of meaning in terms of the original culture. Honey is one of them.

For the San, it was a highly valued food. But there was nothing 'ordinary' about it. In the first place, bees' nests could—unusually—be owned by individuals and inherited along with living places. In their semiarid environment the /Xam lived largely near to, but not at, waterholes.[30] Bleek and Lloyd's /Xam informant //Kabbo explained this point: 'Our grandfather dying left our father his old houses; that our father might dwell at them. Our father did take, [*sic*] his (our grandfather's) honey's hole' (L.II.14.1363). //Kabbo is here referring to bees' nests that, out on the dry, open plains of the interior, were often underground rather than in precipices. Similarly, in the southeastern mountains where Orpen recorded Qing's tales, 'honey was greatly esteemed and when the nests of wild bees were discovered they were marked by the finders and subsequently regarded as their own property, which could be handed on to their children' (How 1962:46). In the mountains, bees' nests are in cliffs, as is the one in Qing's myths. In the Kalahari Desert today, where there are for the most part no cliffs, a person who finds honey, probably in a tree, marks it with a bunch of grass or some other means. Comparably, Cagn marks the honey with 'his sticks'. In an act of defiance, Qüuisi throws them away. This is a nugget, a deeply serious matter, not a minor misdemeanour. In the 1950s, Lorna Marshall (1960:336) heard that a Ju/'hoan man was killed because he took honey that belonged to someone else.

So far, the surface narratives of both tales seem restricted to infringements of social relationships and the restoration of a desired food to its rightful owners. Thus understood, performances of the tales could be said to have reinforced the rules of sharing that are fundamental to San life. In this sense, the myths were didactic. But there is more to it. That the dispute between Cagn and the eagle concerned honey rather than some other resource, such as plant foods, takes the narrative from a general affirmation of sharing to a specific religious, or spiritual, context. In the next chapter, I consider honey in mythic and ritual contexts that bring out the more esoteric significances that gave it its extraordinary status as a richly resonant nugget. As we shall see, honey is noted for its supernatural potency and thus association with San shamans.

When the supernatural components of Qing's stories are taken into account, further levels of significance become apparent. While sharing of food does indeed underwrite the ideological egalitarianism of San groups, egalitarianism may, as we shall see, be subverted in ways other than privileged access to material resources. San egalitarianism (and no doubt other egalitarianisms elsewhere in the world) is not monolithic but rather situational. Indeed, the essence of what I call subverted San egalitarianism is woven into Qing's tales. The tales seem to reinforce sharing and the egalitarianism that sharing implies, but, at the same time, they promote and mask the reality of social differentiation: some people, like Cagn, have supernatural abilities, while others do not. Once that difference is acknowledged, it becomes apparent that Qing's honey tales are more about supernatural potency and the benefits of possessing it than about acquiring a delicacy.

Charms and Potency

In both Maloti myths, Cagn achieves his ends by resorting to his charms. In his first tale Qing said: 'Presently, he thought of his charms, and took some from his belt, and caused them to go to Cogaz to ask advice. And Cogaz sent word back by means of the charms that he was to make water to run down the rock, and he would find himself able to come down'. In the second tale, he said: 'And Cagn began to bethink him of his charms, and he sent to ask Cogaz for advice through his charms'. What are these 'charms'?

The /Xam spoke of a number of things that Bleek and Lloyd designated 'charms' (*≠kan*). They constitute another nugget. In material form, /Xam charms included a piece of a hartebeest's hoof (the hartebeest was one of /Kaggen's 'things'), baboon hair, and aromatic herbs, the species of which were probably variable.[31] These charms were used to protect children, to secure good hunting, and to ward off illness and danger.

But charms were not exclusively good luck talismans. In one passage, Lloyd translated the /Xam word *//ke:n* as 'charm' and also as 'enchantment' (Bleek 1935a:33). *//Ke:n* usually denotes supernatural potency. Bleek and Lloyd also translated it as 'magic', 'sorcery', and 'bewitch'.[32] At least partial synonyms include *!gi:* (as in *!gi:xa,* a person filled with potency) and */ko:öde.*[33] We need to sort out the meanings of these three words. Are they synonyms? This is a difficult question that has tripped up some writers. The only way we can answer it today is to examine the contexts in which /Xam people used the words. I give two examples. The first is a passage that deals with the death of a /Xam shaman. Diä!kwain said that 'our mothers' had taught them about potency. He continued:

> [The deceased] takes the magic power [*/ko:öde*], he shoots it back to the place where people are. For the people are those whom he wants to take away with his sorcery [*//ke:n*], for he thought of them while he was among men . . . a sorcerer is a being who[,] when he dies, wishes to fall heavily taking his sorcery [*!gi:*]. (Bleek 1935a: 28–29)

As used here, the three words seem to be interchangeable, whatever subtle connotations they may have had in some contexts. In another context, Diä!kwain used two of the words when he was speaking about locusts, a desired food that was drawn into the web of mystical things. He said that shamans control the appearance of locusts: they 'charm them with magical doings [*//ke:n-ka didi*] . . . locusts only go about because of magical doings [*/ko:öde*]' (Bleek 1935a:10–11). In this sentence, *//k:en* and */ko:öde* appear to be synonyms. In the following pages we encounter Lloyd's word 'sorcery' a number of times. In each case, the speakers used one of the three words we have discussed: *!gi:*, *//k:en,* and */ko:öde.*

Bleek and Lloyd's 'sorcery', then, is the manipulation of supernatural potency for a variety of ends, some good, some malign. Whichever word people used, potency was an invisible essence that permeated the San world and that people struggled to control.

San charms were in some way allied to this potency, although they probably had multiple uses. Certainly, a link between some of them and the trance dance is clear. Commenting on a rock painting depicting human figures with antelope heads, Qing (in Orpen's translation) said:

> Cagn gave us the song of this dance, and told us to dance it, and people would die from it, and he would give charms to raise them again. It is a circular dance of men and women, following each other, and it is danced all night. Some fall down; some become as if mad and sick; blood runs from the noses of others whose charms are weak, and they eat charm medicine, in which there is burnt snake powder. (Orpen 1874:10)

Alhough we do not have Qing's San word, 'charms' seems here to be a substance, although 'whose charms are weak' may refer to inadequate potency in a shaman—in quality or quantity. Qing's mention of 'burnt snake powder' as a component of 'charm medicine' leads to further associations. The /Xam said that the Mantis can become a snake (Bleek 1924:11), and the Ju/'hoansi have a mamba medicine dance song (Marshall 1962:249). A Ju/'hoan man may be said to possess snake (mamba) medicine (England 1968:431). A small group of San, who until recently were living as a recognised community near Lake Chrissie in present-day Mpumalanga Province of South Africa, reported that San 'doctors' plunged into a pool to catch a large snake and then danced with snake skins (Potgieter 1955:30). Waterholes were significant places (Chapter 3). In one of Orpen's myths snakes are associated with rain and water (Orpen 1874:7). Similarly, the /Xam spoke of certain snakes as 'the rain's things' (L.V.6.4384', 4385'). All in all, it seems that 'snake powder' was probably believed to be imbued with potency. In addition to 'snake powder', the southern San used aromatic herbs that were also believed to be imbued with potency (Bleek 1936:144–60; Hollmann 2004).

Potency was, in turn, associated with the music of a dance. The Kalahari San believe that the potency of the trance dance resides, at least in part, in the various 'medicine songs' that come ultimately from god. Biesele (1993:67–68) met a Ju/'hoan woman named Beh who told how, one day out in the veld, she saw a herd of galloping giraffes running before a thunderstorm. 'The rolling beat of their hooves grew louder and mingled in her head with the sound of sudden rain. Suddenly a song she had never heard before came to her, and she began to sing. G//aoan (the great god) told her it was a medicine song' (Biesele 1993:67).

Beh was still alive in 1991 (Biesele, pers. comm.). Similarly, Qing said: 'Cagn gave us the *song* of this dance' (emphasis added). Shamans who become ecstatic and eventually fall unconscious in trance are said to have 'died': they are visiting the spiritual realm, as all people do when they die (for example, Katz 1982:115–16, 214–16). Hence Qing's words: 'People would die from [the dance], and [Cagn] would give charms to raise them again' (Orpen 1874:10).

Supernatural Contact and Escape

Without access to the San words that Qing used, we cannot be certain that only one was translated as 'charms'. But an important point is clear enough. A link, perhaps not always present, between charms and the multifaceted shamanic enterprise is evident in both of Qing's myths. Cagn's charms enable him to establish out-of-body contact with Cogaz and to find out how to escape from his dilemmas. This sort of supernatural contact is still part of San shamanism. A Kalahari shaman explained that he can send messages along 'threads of light', iridescent lines that people routinely perceive when they are in trance. He added: 'It is similar to what you people call a telephone' (Katz, Biesele, & St. Denis 1997:108–09; Keeney 2003:42;). This 'telephone' experience is probably what Qing was referring to when he said that 'Cagn . . . sent to ask Cogaz for advice through his charms'.[34]

Part of the advice Cagn received was 'to make water to run down the rock'. He would then find himself able to come down. There is more to this nugget than at first meets the eye. A clue comes from the Bleek and Lloyd Archive. 'To make water' is Lloyd's translation of *//khū,* 'to urinate'; it is, or was, of course a common English euphemism (Bleek 1956:577). We do not know what San word Qing used, but it seems clear that Orpen was using the same euphemism as did Lloyd.

In the context of San thought and in this narrative in particular, 'urinate' makes sense. It is another nugget. Urine had significance for the San beyond its literal meaning of human waste. For example, when /Xam hunters who have wounded an eland returned home, 'they made water upon the eland' to ensure success the following day (L.V.6.4413). The Ju/'hoansi say that (the lesser god) //Gauwa's urine carries potency, and he places it in shamans' tortoise shells along with the urine of a supernatural giraffe (Marshall 1999:57). A coal dropped into the shell produces a smoke that is believed to induce an altered state of consciousness in the dancers, *!kia* in the Ju/'hoan language. Working with the !Kõ San, the linguist Anthony Traill (2009:117) learned that 'a person's urine is potent medicine'.

It thus appears that Cagn was using his potent urine to escape from his dilemma. 'To make water' turns out to mean much more than simply 'urinate': as a nugget, it ties in with other words and phrases in the myth that are all parts of the pattern of San shamanism. A potent substance in the hands of a shaman effects escape from a difficult situation. In Qing's tale, the Mantis's escape is sealed when the eagle is killed by hail.

Hail

Hail is yet another nugget. Anyone who has been caught in a Maloti-Drakensberg hailstorm knows that it can be lethal. Indeed, in the land of the /Xam, /Han≠kass'o's second wife was killed by a hailstorm (Bank 2006:281). But, for the San, hail was not merely a natural phenomenon. Hail, rain, and lightning were believed to be under the control of !Khwa, in being or rain-animal form. Hail was especially associated with dangerous thunderstorms, known as the 'male rain'. /Han≠kass'o explained that an 'angry' rain 'attacks the hut angrily, and the hail beats down on us breaking down the huts' (Bleek 1933:299). In a note about the porcupine, Diä!kwain said that his parents taught them that they should not interfere with the 'hole' (*//khauru*) on the back of a porcupine; if they did, hail would fall and kill them (L.V.6.4387'). In the Ju/'hoan story of G!kon//'amdima, the woman invokes the rain and causes hail to fall on her pursuers (Biesele 1993:168).

But hail in Qing's tale is evidently under the control of Cagn: he can control !Khwa. The mystical associations of hail as a supernaturally wielded weapon would have been obvious to San listeners and be at one with the general tenor of Qing's tale.

Foundational Beliefs

As I have pointed out, in some authentic San contexts these tales about Cagn and the honey were probably dramatised and acted out by the narrator, much to the amusement of all present. For the San, fun and even ridiculing the trickster are not incompatible with profound social issues. But, as is so often the case, there is more to it. We can now see that the central idea of Qing's two tales is much more than a trickster's entertaining antics and the underlining of the San sharing ethic.

In both tales Cagn finds himself in trouble with a malevolent being but is able to use his special abilities to extricate himself and to bring honey home to his people. As we shall see in the next chapter, Qing's two tales are—centrally, though not exclusively—about the possession and activation of potency. Bringing honey home is important for Cagn's whole family. Its potency will provide them with healing and protection against the 'arrows of sickness' that malevolent shamans and spirits of the dead shoot at people. Those who possess potency, the shamans, are able to protect their communities. The myths become intelligible when we realise that Cagn himself is a shaman who can travel underground, transmit thoughts via his charms, and unleash hail.

This is no small point. Shamans and their access to the supernatural realm are at the hub of San life. Everything from economic activities (for example, hunting antelope and gathering locusts) to protection from supernatural forces and agents entails bridging the chasm that could all too easily open up between material and spiritual cosmological realms. In this way, Qing's two tales underline the shamans' social status in San communities. Anyone who withholds honey/potency and so disrupts the equilibrium of life, as the eagle did, is dangerously antisocial.

With these fundamental understandings of the Mantis and the importance of potency in mind we can turn to the more elaborate /Xam narratives that Bleek and Lloyd recorded. They not only reflect what Bleek called 'the general character' of the two corpora of nineteenth-century San myths: they also raise social concerns and issues of potency similar to those embedded in Qing's myths, although in different ways. San unity in diversity becomes clearer.

CHAPTER THREE

The Mantis Makes an Eland

He called the eland, and the eland got up, and came forth, while the ground sounded.
(/Han≠kass'o, L.VIII.6.6525)

'The Mantis Makes an Eland' is the title that Dorothea Bleek gave to a series of complex /Xam tales about the creation and death of the first eland antelope (*Tragelaphus oryx*). Her father and aunt had recorded them in the 1870s; she prepared them for publication in the early 1920s.

Varying in length, performances of sections of the narratives were given by four /Xam men: /Han≠kass'o, Diä!kwain, //Kabbo, and ≠Kasin. In total we have five transcriptions; however, in keeping with the fluidity of San mythology, none is what could be seen as a full, all-inclusive narrative. Nor was any a transcription of an authentic San performance triggered by purely San concerns in a San social context. A Victorian home and garden and transcribers who were interested in certain aspects of San folklore and life—and baffled by others—must have been factors that exercised some control over what the informants said.

In different, although comparably inauthentic, circumstances in the southeast mountains, also in the 1870s, Joseph Orpen obtained a variant of the tale from Qing (Orpen 1874:3–5). It differs in some respects from the /Xam texts, but it follows the sequence of creation and death of the eland; it also adds an episode of re-creation (Wessels 2014). An independent and again somewhat different variant was later heard in Namaqualand at the beginning of the twentieth century (Curlé 1913). Most recently, a variant was collected in the second half of the twentieth century from the northern Kalahari San. As in the other texts, the trickster creates an eland, and the animal is prematurely killed (Heikkinen 2011:59–62, 168–69).

Myth and Meaning: San-Bushman Folklore in Global Context by J. D. Lewis-Williams, 75–96

Together, these records show that the eland creation myth *in its broad outline* was not short lived, nor was it limited to a narrow geographical area. 'The Mantis Makes an Eland' and the subsequent death of the first eland clearly invoked a set of concepts that were important, not just in the judgement of researchers but to the San themselves across southern Africa. This southern African instance recalls the ancient Sumerian Epic of Gilgamesh (George 1999). It, too, recurs over a wide geographical area and, as additional clay tablets are discovered, researchers piece together a 'complete' narrative. Nevertheless, the reconstruction of a 'complete narrative' by splicing together the various recorded segments of the eland myth would be pointless: no such thing ever existed. There was no Platonic 'true' version comparable to a canonical form preserved in a sacred, written text. Christian theologians try to synthesise the four Gospels; the San would see no point in doing so. Rather, they selectively drew on a rich source of raw material for specific, but ephemeral, performances.

Megan Biesele has considered the Western notion of a 'correct' story in relation to multiple San variants. She notes that some San narrators insist that they are faithfully repeating an old story 'when what they are doing is actually telling it in their own words *in a way that reflects the performance situation of the moment*' (Biesele 1993:67, emphasis added; 1986). For the San, the episodes and variants existed outside linear time, and arranging all of them in a 'correct' sequence would have been largely meaningless.

Much academic work on myth in general has none the less focused on the structures of 'complete' narratives and has sought recurring patterns in a number of variants.[1] Some scholars believe the 'message' of a myth to be hidden principally in the repeated syntagmatic or paradigmatic narrative structures of those variants. I have not found this notion a sufficient approach to San myths. In some instances, emphasis on the fundamental concept of a narrative with a beginning, a middle, and an end with an implied moral is unhelpful in that it circumscribes understanding. To be sure, I acknowledge that the narrative of a myth can be significant—as I showed in Chapter 2. After all, one cannot have a story without a series of successive events in which the end differs from the beginning. That is implicit in the very notion of time. But narrative structures, shorn of the specificities implied by nuggets, may be inadvertently manipulated by researchers to convey 'messages' that accord with their own morals and theories of myth.

Even if some repeated structures can, as I showed in the previous chapter, indeed be found in San myths,[2] structuralist methods nevertheless seem to start, as it were, at the wrong end. Emphasis on overall structures tends to take the narratives out of their San contexts: the structures become universal templates that, without their specific ethnographic settings, carry few, if any, inherent meanings. A narrative structure may be compared with a quadrupedal armature on which many different clay animals may be fashioned. The ethnographic approach that I adopt tries to combat this problem. I argue that it is better to start with detailed investigations of what are essentially San elements embedded in the narratives. In the previous chapter honey and charms were two of them. Then, in the next analytical step, it becomes apparent that these elements interlock and *together* suggest what the tale might, allowing for different emphases in different performances, have meant to some San in some social and religious contexts.

These are some of the reasons why this book did not start with a discussion of 'theories of myth', such as those devised by Malinowski, Jung, Freud, Lévi-Strauss, or Campbell. I tend to move away from grand 'theories of myth' and try not to allow them to guide my thinking, as I formerly did (Lewis-Williams 1972). The eland myths are a case in point. If we attend to specific nuggets, it becomes apparent that the episodes do not constitute a simple 'creation story', despite Dorothea Bleek's title. They are not comparable to Rudyard Kipling's *Just So Stories*, which were immensely popular at the time when she published the eland creation story in *The Mantis and His Friends* (Bleek 1924). True, the narrative seems to teach, first, that the Mantis created the original eland and, second, that people should not hunt eland without first asking his permission. But why does the myth do this? Does it merely satisfy some supposedly innate desire in all human beings to explain things (a highly problematic concept)? If so, why did the eland merit such an explanation and not other creatures that feature prominently in San life and thought, such as lions? If the myth's explanation of the eland's origin was merely intended to satisfy curious humanity, why was it so widely known and in so many variants? Surely, the myth must have dug deeper into San thought than curiosity as to where a certain animal came from. If we adopt an aetiological approach to myth, we must ask why the origin of these particular things is explained and not a myriad others? Indeed, worldwide, if we list the things that a people's origin myths explain, we shall begin to form a better idea of what those myths are 'doing'. Having perused our list, we need to ask: who benefits from the explanations of these particular things?

In short, to see the eland myth as merely a bald explanation of the origin of a species of antelope is unpersuasive. By situating the narrative in a web of conceptually interlocking beliefs and social interactions we move closer to the widespread power that the tale seems to have exercised for many years in San communities across southern Africa.

/Han≠kass'o and His Variant

Dorothea Bleek published what she called two 'versions' of the eland narrative (Bleek 1924:1–9). Her father's and her aunt's manuscripts, however, show that her so-called First Version is her own undeclared combination of performances given on a number of separate occasions by two informants, //Kabbo and Diä!kwain. Like Orpen in his work with Qing, she tried to make the tales 'consecutive'. In her original manuscript, Lloyd titled Diä!kwain's performance 'How the Ichneumon Discovered What the Mantis Did with the Honey' (L.V.1.3608). She presumably heard a summary version of the tale, devised her title, and then asked for it to be dictated in full so that she could note it down word by word. As she first understood it, the story seemed to emphasise possession of a desired food rather than the creation of an antelope. Only later did Dorothea focus her title on the creation of the eland: 'The Mantis Makes an Eland (First Version)'.

By contrast, Dorothea's 'Second Version' is a slightly shortened form of a single narration that /Han≠kass'o gave to Lucy Lloyd during the course of nine sessions spread over fifteen days in 1878 (Figure 3.1). Dorothea translated it in December 1914. It is principally this performance of the myth that I discuss.

To understand /Han≠kass'o's interest in the tale we need to know something about him and his concern with things mystical. His life epitomises the harsh circumstances that the San endured on the fringes of the Cape Colony during the nineteenth century. He was known as a 'Flat Bushman', after an area that took its name from the extensive plains near the present town Vanwyksvlei (Deacon 1996). Like other /Xam informants, he had worked intermittently for Dutch-speaking settlers, the Boers. His Dutch name was Klein Jantje. His father had died before the Boers reached that part of the country (L.VIII.1.6052–3), although the disruption and suffering they caused would have preceded their actual settling in the area.

He said that //Kunn, a rain-shaman who was related to his mother, 'loved [him] dearly'. He observed the effects of //Kunn's rainmaking: 'I have seen the rain he caused to fall. . . . His rain came streaming from out of the north, because he was from that part' (Bleek 1933:387). Rather boldly, and ignoring his mother's admonitions, he imitated //Kunn's rain-making technique by

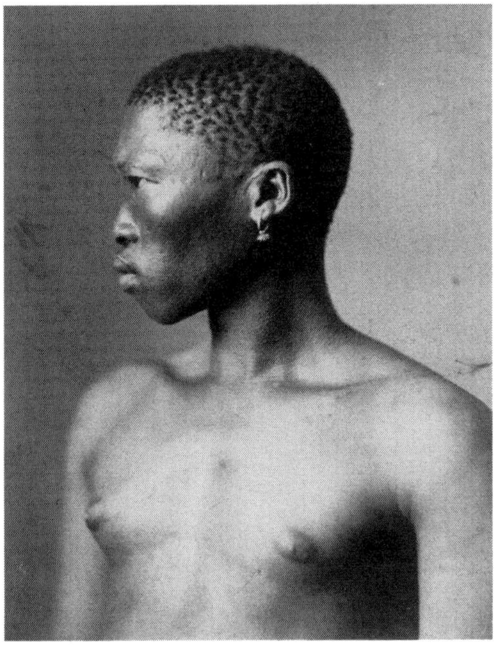

FIGURE 3.1 A nineteenth-century photograph of /Han≠kass'o

playing a musical instrument known as a *goura*, but with disastrous effect: a severe lightning storm ensued.[3] /Han≠kass'o did not join the Bleek family when he was a prisoner in the Cape Town Breakwater Prison, as Diä!kwain did, but he must have known about the work that Bleek and Lloyd were doing with /Xam people: his wife was a daughter of //Kabbo, one of the Bleek family's most prolific informants (Chapter 4), and he had presumably heard about the Bleek family from her and her father. After he had gone home at the completion of his prison sentence Lucy Lloyd invited him to return to Cape Town. Wilhelm Bleek had died in 1875, and she was continuing with their work. At about 30 years of age, /Han≠kass'o set out from his home with his wife and an infant. Tragically, the child and his wife both died on the journey, she having been weakened by an attack on her by a policeman (Bleek & Bleek 1909:41). He arrived in Cape Town alone. He was with the Bleek family from January 10, 1878, to December, 1879. During that time he pined for his young son !Hu!hun, whom he had left with friends at home. Researchers have cited /Han≠kass'o as the most fluent of the /Xam informants in story-telling.[4]

The Overall Narrative

I now show how themes of the two Maloti honey myths that I discussed in Chapter 2 are woven into the /Xam eland creation narratives. This interweaving of themes in apparently different tales from widely separated areas again shows that it would be wrong to suppose that San myths are single, direct narratives, each with its own distinct meaning. Throughout this book, I argue that the bare narratives, although not unimportant (it was, after all, the Mantis and no one else who made the eland in each account), were the framework for a number of themes. To those unfamiliar with the esoteric significances of San nuggets, what appear to be peripheral themes may in fact be closely related to the main story.

Space does not permit a full verbatim presentation of even one of the performances of the /Xam eland creation myth, let alone discussion of all the themes embedded in them (Bleek 1924:2–9). I do not pretend to have exhausted all the meanings enshrined in the tales. In his study of /Xam San myths, Roger Hewitt (2008:174, 176) lists 'three main headings' and ten 'main events'. For the purposes of this book I distinguish four major episodes, each of which comprises a number of elements. Other divisions are no doubt possible.

The Mantis Makes an Eland

- The Mantis creates the first eland by placing a shoe, or sandal, in a waterhole. //Kabbo said it was the Mantis's son-in-law /Kwammang-a's shoe; other versions say it was the Mantis's own shoe.
- The Mantis should be bringing home honey, and the people at home wonder what he is doing with it.
- He is in fact feeding the honey to the eland and rubbing it, mixed with water, on the eland. He left the eland in the waterhole for three nights, during which it grew as large as an ox.

Subversive Elements in the Family

- To find out what is happening to the honey /Kwammang-a tells his son, the Ichneumon, to spy on the Mantis. The Meerkats then go to the eland and kill it without the Mantis's permission.
- While still collecting honey, the Mantis finds that it is 'dry', an omen that tragedy has come upon his house.

- Finding blood, the Mantis realises that his 'child' has been killed.
- He weeps and follows the dying eland's tracks.

Conflict

- The Mantis finds the Meerkats who have killed his eland. He tells them that they should have waited for him to tell them that they could kill the eland.
- He shoots arrows at them, but they fly back at him and he has to dodge them.
- He tells the eland's gall, which has been placed on a bush, that he wants to pierce it. The gall agrees and tells him to spring into the blackness that will ensue. He pierces the eland's gall and leaps into the blackness.

Resolution

- The Mantis creates the moon from a shoe or, depending on the variant, an ostrich feather that he sucked and used to wipe the eland's gall from his eyes. He does this to enable him to see in the darkness and find his way home.
- He then uses his powers to cause the eland meat and the tree on which it and the Meerkats' belongings are hanging to fly through the air to his camp. The honey/eland thus comes home.

In this chapter I deal with the first two sections. Chapter 4 discusses 'Conflict' and 'Resolution'.

Primal Time

Where and when were these events supposed to have taken place? The Bleek family's informants sometimes alluded to a Primal Time, or First Order, that they said was inhabited by people of the Early Race. It was in this 'removed', rather than just temporally remote, period that they said many of the myths took place. It was then that people were animals and animals were people. But the two periods were not isolated from each other. The Primal Time spilled over into the present time. Mathias Guenther (1999:66) has perceptively and succinctly expressed this idea: 'This confounding of past and present, and myth and reality, confers on the present order an abiding aura of ambiguity.' This ambiguity is important. It expresses itself in numerous, often subtle, ways, not only in myths. For instance, the empathetic bond between a San hunter and his prey warrants a series of complex rituals that invoke supernatural relationships; in addition, traces of humanity can still be discerned in some animals.[5]

Guenther also sees an important link between the Primal Time and trance dancers:

> Like the beings of the First Order in mythic time, the dancers of today may shift their ontological state, crossing back and forth between boundaries that separate them from spirit and animal beings, from life and death. At one phase of the dance, before the moment of collapse, they stand on the threshold of both realms, in full view of the spectators and participants, who witness the dancer's intense experience of dissociation. (1999:191)

Note that myth is not something separate from the inter-realm contact achieved by trance dancers. There are significant parallels between myth and dance.

In practice, however, it is not always easy to tell if a /Xam tale relates to the Primal Time or to the present (Brown 1998:54). I suspect that in many instances the Bleek family's narrators did not specify the Primal Time location of a myth because they assumed that their auditors would realise this. The eland myths, however, have all the features of the Primal Time.

The Mantis's Family

To understand /Xam myths, we need to begin with the kinship relations of the principal actors. As in many myths throughout the world, kinship is central to understanding. However, the ways in which people regard different categories of relatives differ from society to society. To discern what was going on in /Xam myths we need to know what certain specific /Xam familial relationships and tensions meant to the people. The Mantis's family is in some ways a microcosm of /Xam society, and the myth gives us insight into life in a small /Xam camp. Nevertheless, the membership of the Mantis's family is, like the tales themselves, not rigidly defined. The members vary from tale to tale, although those that appear in the eland episodes seem to be fundamental (Figure 3.2).[6]

I discussed the Mantis and his status as an ambivalent trickster and original shaman in Chapter 2. We come now to the other members of his family. He is married to the Dassie (hyrax, rock rabbit). Wilhelm Bleek tentatively

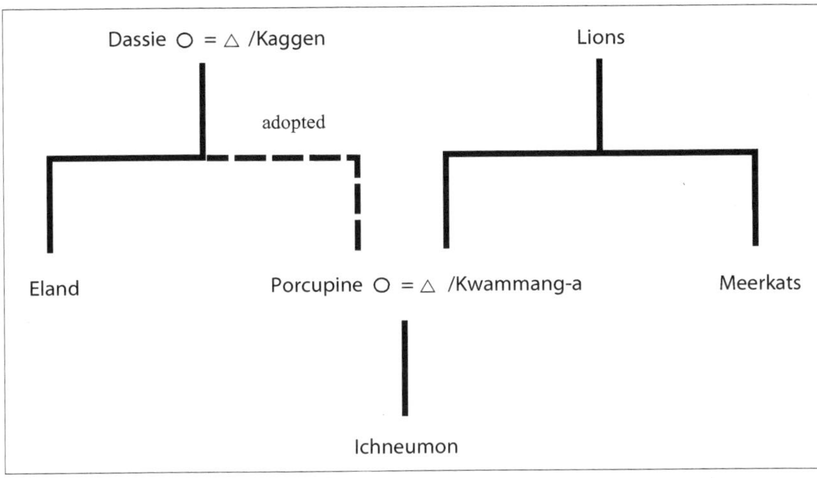

FIGURE 3.2 Relationships within the Mantis's family

pointed out that Coti, the Maloti name that Orpen recorded, 'may be identical with the beginning of one of the names given to us for the Mantis' wife */húnntu !(k)att !(k)atten*, the first syllable of which word */hunn* indicates a "dasse" ("dassie"); but this is not certain' (Bleek 1874:11).

In the /Xam creation myths, the eland is spoken of as /Kaggen's son. In the Maloti variant, Coti gives birth to the first eland (Orpen 1874:3). In the /Xam tale of the //Khwai-hem (a monster), the Mantis is said to have another son. He is known as 'the young Mantis' (Bleek 1924:38), but he does not appear elsewhere. The eland is the Mantis's most significant offspring. It is the largest African antelope; an adult bull weighs up to 940 kg (Skinner & Chimimba 2005:638–42). It is also the most tractable African antelope. Fleet hunters are able to run down an eland and drive it back to their camp before killing it.[7] San beliefs, myths, and rituals recorded in both the nineteenth and twentieth centuries show that this antelope, although valued for its meat, occupied a unique status in San thought. Lorna Marshall lists the symbolic associations of the eland as she inferred them from the Kalahari Ju/'hoansi in the 1950s: meat, fat, health, strength, rain, fertility, plenty, unaggressive behaviour, and general well-being—all life-giving things (Marshall 1999:82, 195, 268). Moreover, regional quantitative surveys conducted in the 1960s and early 1970s confirmed a long-held impression that the eland was the most frequently depicted animal among the paintings of the mountainous areas of southern Africa (on which I concentrate in Chapter 7) and also among the rock engravings of the central plateau.[8] In sum, there can be little doubt that, for the San generally, the eland was rich in significances.[9] When speaking to Megan Biesele and me in the 1970s, a Ju/'hoan man summed up the importance of the eland by saying that 'other animals are like servants to the eland' (Lewis-Williams & Biesele 1978:117.) His remark brings to mind an early Western missionary's observation that, among the antelope he encountered, the eland surpassed 'all others in beauty, and [moved] amongst them like a king' (Arbousset & Daumas 1846:65).

The Mantis's adopted daughter is the Porcupine. She escaped from her frightening father, the monster known as //Khwai-hem. Lloyd did not translate this name, but Dorothea Bleek gave it as 'the All-Devourer' (McGranaghan 2014a). In a long tale told by //Kabbo, //Khwai-hem eats the Mantis but, in the end, is himself killed (Bleek 1924:34–40). //Khwai-hem behaves like a veld fire consuming everything in his path (Bleek 1924: 34–40), and this is probably the reason for Dorothea's 'translation'. It does, however, seem that his name is more specific. *//Khwai* means 'quiver', and *hem* means to 'devour'. In fact, Dorothea Bleek's *Bushman Dictionary* explicitly suggests that the name may mean 'quiver devourer' (Bleek 1956:59, 60). If this translation is correct, as I believe it to be, the tale takes its place alongside others that deal with frustrated food-getting and the restoration thereof (Chapter 6; Lewis-Williams 2013a). The decidedly unpleasant //Khwai-hem is, moreover, referred to as 'cousin', and this may categorise him as an affine, of whom more in a moment (Bleek 1924:35, 37). In a further extension, the creation of this monster in /Xam folklore was probably part of the /Xam's engagement with other peoples, pastoralist Korana and European settlers, who were taking over the land. The /Xam seem to have spoken of the invaders in terms of traditional monster categories.[10] In his discussion of 'monstrosity', Mark McGranaghan (2014a) shows that the /Xam folklore that the Bleek family encountered was not static and atemporal. After all, in the land of the /Xam the tales were being told against a background of violent disruption. Myths should not be removed from their place in history.

The Porcupine is married to /Kwammang-a, but she lives with the Mantis and the Dassie. Lloyd noted that she could not 'yet understand' why the Porcupine lived there (L.II.33.3046'). Her 'yet' suggests that she hoped to find out more in due course. Many years later the answer came to researchers from the Kalahari San. It is still customary in many San communities for a son-in-law to live with the bride's family rendering bride service for a few years, at least until he is able to demonstrate his ability to care for his wife. The service a man gives is primarily hunting, and, among the Ju/'hoansi, it may last long enough for three children to be born. The couple may then go to live with the man's family if they so wish (Marshall 1976:168–72). Having married the Porcupine, /Kwammang-a is performing bride service in /Kaggen's family.

The Porcupine and /Kwammang-a's son is the Ichneumon, a small furry creature similar to a mongoose and therefore akin to a meerkat. In the tales, he occasionally lectures the Mantis on how he should behave, something that the Porcupine and /Kwammang-a never do. As a grandson, the Ichneumon has an easy 'joking relationship' with the Mantis (Marshall 1976:204–08).

As a result of the Porcupine's marriage to /Kwammang-a, the Mantis is affinally related to his son-in-law's family, the Meerkats—they are not his blood relations. As is not uncommon in San communities, the Mantis has an uneasy relationship with his in-laws. Certainly, this is clear in the tale we are discussing. By asking the Ichneumon to spy on the Mantis, /Kwammang-a, an 'interloper', is ultimately responsible for the Meerkats' illicit killing of the eland.

Conceptual Groups

There are other parallels, not just kinship relations, that hold the Mantis's extended family together and, at the same time, distinguish one segment from another. All the characters I have mentioned fall into two conceptual groups.[11] These groups and their respective significances play a key role in the generation of the narrative.

The Mantis's nuclear family, the first conceptual group, comprises the Mantis himself; his wife, the Dassie; and the adopted Porcupine. At first they may seem a heterogeneous group of creatures, but we need to ask if they have anything in common in San thought. We then find that they are all associated in various ways with honey and another important substance—fat. Indeed, fat and honey are related substances in San thought. Uniquely, both components of the dyad can be eaten and drunk. In a discussion of San symbolism, Biesele (1993:86) calls them 'two very important foods'. I return to the relationship between fat and honey in a moment.

Historically, the San valued fat. As one early observer put it, the San were 'extravagantly fond of fat' (Steedman 1835:1:150; see also Bleek 1935a:238). They still are today. But there is more to it. Briefly, dassies and porcupines are both creatures known for the large amount of fat that they possess. Moreover, the dassie is an animal that lives in rocky cliffs, which, in turn, the /Xam associated with bees and honey. /Han≠kass'o was explicit on this point: 'The Dasses lived with the Bees' (Bleek 1924: 47). Far to the north, among the Ju/'hoansi, the principal deity's wife is still known as 'Mother of the Bees' (Marshall 1962:226).

Furthermore, the eland antelope is considered to have more fat than any other creature. This is especially true of the male. The association of the eland with fat was not limited to the nineteenth-century /Xam. Today the Ju/'hoansi speak lyrically of the fat to be had from, especially, an old bull eland. One man said that the fat around the heart of an old bull is so great 'that when you have melted it down, you have to make a container of the whole eland skin' (Lewis-Williams 1981:50).

The closeness of the eland's association with fat may be one reason why the linguistically and geographically disparate San groups retained a myth about its creation. In their eyes, it is a special animal, and fat is a powerful substance. For instance, Ju/'hoan shamans place eland fat, among other substances, in their smoking tortoise shells to induce trance at a dance (Biesele 1993:105). Indeed, fat is important in numerous San rituals, including girls' puberty and marriage rites (Lewis-Williams 1981:41–54, 69–74). Of a girl at puberty, a Ju/'hoan woman dreamily murmured:

> The Eland Bull dance is danced because the eland is a good thing and has much fat. And the girl is also a good thing and she is all fat. . . . They do the Eland Bull dance so that she will be well. She will be beautiful; that she won't be thin; so that if there is hunger, she won't be very hungry and she won't be terribly thirsty and she will be peaceful. That all will go well with the land and that rain will fall. (Lewis-Williams 1981:48, 49, 50)

This association of the eland with fat leads back to the link between fat and honey. The Mantis himself, an insect that hunts, is the provider of both fat and honey, as the episodes I am discussing make clear. Hunting and honey gathering were both men's tasks and highly valued.

All in all, the eland was said to be the Mantis's favourite animal:

> The Hartebeest and the Eland are things of the Mantis; therefore they have magic power. . . . People say that the Mantis first made the Eland; the hartebeest was the one whom he made after the death of his Eland. . . . The Mantis does not love us, if we kill an eland. (Bleek 1924:10–12)

Fat and honey, themselves related, united the Mantis's nuclear family.

The second group, the Mantis's affines, contrasts conceptually with his nuclear family. They comprise carnivores—lions in the older generation (lions feature prominently in other narratives) and meerkats (which are also insectivorous) in the younger (Lewis-Williams 1997). /Kwammang-a's origin in this group explains why his and the Porcupine's son is the Ichneumon, a small carnivore seemingly out of place in the Mantis's fat-and-honey-centred family. /Kwammang-a is the only member of the Mantis's family that does not have an explicit animal identity: it is as if his origin as a carnivore or 'pawed creature'—a dangerous stranger in San parlance—is being suppressed. It is this relationship that underlies the fight with the Meerkats that takes place after they have killed the eland (Chapter 4).

Already we can see a connection between Qing's two tales and the /Xam eland creation narrative: honey. Later we shall see that Cagn's fight with the eagle and its attempt to kill him when he was 'in the form of a large bull eland' (Chapter 2) echoes the fight with the Meerkats. The eagle is a carnivorous, avian hunter and so parallels the Mantis's 'pawed' affines in the /Xam narratives. The bringing home of honey and conflict in the spirit realm emerge as central ideas in both tales.

The Creation of the First Eland

The Mantis makes the first eland out of a shoe that belongs to either his son-in-law, /Kwammang-a, or to himself, depending on the emphasis of any given performance of the myth. //Kabbo's performance suggests that the Mantis, as trickster, mischievously takes the shoe without /Kwammang-a's knowledge, while /Han≠kass'o implied that /Kwammang-a had discarded the shoe. This difference may perhaps be tied to the narrator's desire to emphasise or downplay tensions between affines, that is, between /Kaggen and his son-in-law. /Han≠kass'o, it should be remembered, was //Kabbo's son-in-law.

The shoe itself is a nugget. It was probably a piece of eland hide, the San's preferred (though not exclusive) raw material for making shoes.[12] In San thought, an artefact and the raw material from which it is made share essential qualities. Of the Kalahari G/wi, George Silberbauer (1981:132) writes: 'The act of manufacture would not sunder the link between an artefact and its origins'. In speaking to Lloyd, the /Xam narrators did not bother to mention the species of hide, so obvious was it to them. They would also have been familiar with other instances in their mythology in which dismembered animals are reconstituted from a piece of them (Bleek & Lloyd 1911:3–17, 137–45). Part of a creature becoming the whole creature and an artefact reverting to its original form were motifs in San mythology and thought. We shall encounter them again.

The Mantis places the shoe in a waterhole where there are reeds and where it grows into a large bull eland, 'like an ox' (Bleek 1924:7). In one performance of the creation tale, the eland is said to sleep in the water (L.V.III.6529), although Dorothea Bleek, possibly puzzled by the idea of sleeping in water, omitted this indication of close association between eland and water. The /Xam word for both reeds, plants that grow in water, and arrows is *!nwa:* (Bleek 1956:487); San arrows were, and still are, made in part from reeds. There may be a play on words here: the eland was 'born' among 'arrows' and thus in a hunting context. /Xam hunters had to contend with the Mantis's attempts to contrive his beloved eland's escape, and complex rituals were performed after an eland had been wounded with a poisoned arrow and while the poison was taking effect (Bleek 1932). Hunting an eland is an activity in which realms mingle, one in which a hunter may hear the Mantis's warning cry to his beloved creature (Orpen 1874:3).

Still richer connotations of reeds are found in Diä!kwain's narrative 'The Girl's Story; the Frogs' Story' (Bleek & Lloyd 1911:199–205; cf. Bleek 1933:299). In this tale a girl at puberty kills 'the rain's children' (*!khwa-ꝋpuă*), who live in a waterhole. As a result, a whirlwind carries the girls' people and their belongings to the waterhole, where they become frogs. Wind, and especially whirlwinds, are another nugget. The Ju/'hoansi consider whirlwinds 'a death thing' and a 'fight' (a dangerous concentration of potency) and believe that malevolent spirits of the dead come in whirlwinds.[13] For them, //Gauwa, the lesser god, is in whirlwinds.[14] The /Xam spoke of whirlwinds carrying people and things to waterholes and behaving as if they were beings with volition.[15] Whirlwinds thus interlock with waterholes in the pattern of San thought. When the girls and other things are deposited in the waterhole, the girl's father's arrows and other belongings are transformed: 'The [reed] mats (grew) out by the spring, like the arrows; their things grew out by the spring' (Bleek & Lloyd 1911:205). Everything, it seems, is unstable, in a state of flux.

In a footnote added by a female informant we learn that the characters in 'The Girl's Story; the Frogs' Story' were people of the Early Race. This informant's name was !Kweíten ta //kēn, which probably means 'Person of Potency', *//kēn* being one of the /Xam words for supernatural potency. She was evidently a shaman and would therefore have had personal knowledge of transformation. In Diä!kwain's tale, potency, a girl in puberty (considered potent and therefore a potential danger), whirlwinds, waterholes as places of transition, and origin all come together in a dense congeries that situates the basic narrative.

Creation and Cosmology

The eland's emergence from the water in response to the Mantis's call invokes the three-tiered San cosmos and the implications of what the cosmological realms signify. For the purpose of analysis, the /Xam San's semiarid conceptual world may be seen as biaxial.[16] On the horizontal axis were the camp where relatives and other known people lived and, conceptually if not literally opposite to that, the hunting ground where strangers could be encountered and sometimes dangerous relations with animals were played out. San camps were usually located a mile or so from waterholes so as not to frighten off animals that came to drink (Bleek 1924: unnumbered page). The waterhole was thus an ambivalent meeting place where people could find life-giving water and game animals but also confront potentially dangerous strangers and predators.

The vertical axis seems to have been more defined and important. It comprised spiritual realms above and below the level of daily life. Though by no means exclusively, the lower realm was associated with the dead. Lloyd heard that the dead 'walk along this path; they reach the great hole and they live there' (L.II.6.669', 670'). Above the level of daily life was the realm of god and the spirits. The two axes intersected at waterholes and other places. The San vertical axis was thus akin to the *axis mundi* encountered in numerous societies—a route between the levels of a tiered cosmos.[17] If the river was a montane equivalent to the /Xam waterhole of Qing's tale (Chapter 2), then the 'thing' that held him in the water was a subterranean spirit.

Ascent to the sky on the vertical axis by San shamans sometimes started, to us somewhat paradoxically, by entering a hole in the ground. In an altered state of consciousness (trance, as at a healing dance, or a dream), those who specialised in rain-making (*!khwa-ka !gi:ten*) captured an imaginary rain-animal (*!khwa-ka xoro*) in a waterhole; they then led it out of the waterhole and through the sky to places where rain was needed. There they killed it so that its blood and milk would fall as rain.[18] In a /Xam tale, hunters kill an eland, but, when they put its meat on their fire, it evaporates and they cannot find it among the ashes. Here we see that the eland itself could become the rain and was thus akin to a rain-animal (Lewis-Williams 1981:106–07; 2002a:222–23)—not surprising for an animal that sleeps in a waterhole.

Waterholes were thus mediators, or apertures, between cosmological realms (Lewis-Williams & Pearce 2004a:51–66). In addition, water itself was an interrealm mediator: it falls from above and wells up in waterholes. The /Xam expressed this concept in their single word for both rain and water, *!khwa*. Among the /Xam, newly fallen rain, *!khwa: //ka:n*, was believed to have supernatural potency. It seems in some contexts to have been personified (Lewis-Williams 1981:52). Dorothea Bleek compared *!khwa* with the Mantis, because it was 'something to be respected or feared, as it has power to change people into animals' (Bleek 1929:307). In another linkage in this pattern of San thought, Nicholas England found that the Ju/'hoansi have two high-potency songs constructed in what he took to be a very ancient scale. He called it the Rain-Eland Scale. The Rain Song is sung at healing dances, the Eland Song at girls' puberty rites.[19] One of the reasons for performing the Eland Song at a girl's puberty Eland Bull Dance is that it will bring rain (Lewis-Williams 1981:50). Girls' puberty and shamanic rituals are thus linked by a chain of associations.

This tiered cosmology is fundamental to many San narratives, but we should not think of it too rigidly; it is a researcher's construction inferred from numerous San accounts. In practice, the spiritual realms intermingled with ordinary life. The Mantis, for instance, could be 'by you' wherever you were, even if you did not realise it (B.XXVI.2463'). Traditional Christian cosmology is a parallel: heaven is above and hell below, yet spiritual things are believed to be constantly encountered in daily life.

In terms of this cosmology, the eland is a transitional creature that came, via a waterhole, from the subterranean spiritual realm into the daily world. It is also a transitional creature in another sense: it bridges the male and female categories that are so important in San thought (Biesele 1993). The Ju/'hoansi recognize that the male eland is unique among antelope because it has more fat than the female of the species (Lewis-Williams 1981:50), a curious point that was also noticed by early Western observers to the south (Arbousset & Daumas 1846:45). In this sense, the eland was somewhat androgynous, a point borne out by a rock engraving of an eland with both a male and a female dewlap (Dowson 1988). In addition, and probably consequently, the eland was an *animal de passage* in that it featured in three rites of passage: girls' puberty rituals, boys' first-kill rituals, and marriage rites. In these contexts, it facilitated changes in social status (Biesele 1993:136; Lewis-Williams 1981).

Certainly, the eland was no ordinary antelope: its rich symbolism merited a complex creation myth. In creating the eland, the Mantis was creating not just a favourite animal or even an *animal de passage* but, as we shall now see, the foundation for San contact with the supernatural (Lewis-Williams 1981, 1997).

Bringing Home the Honey

Although the creation of the eland from a shoe is the most prominent narrative component of the performances that Bleek and Lloyd recorded, another theme is woven into the events. From the point of view of our enquiry, this seemingly secondary theme is important. It comprises the relationship between the Mantis, his family, the bringing home of honey, and the part that honey played in the creation of the eland.

Honey is clearly identifiable as part of a subplot. The social setting here is the same as for Qing's two myths (Chapter 2): the Mantis is failing to bring honey home to his family. In the /Xam variant, he is feeding it to the eland that he is creating in a waterhole. He is also anointing the eland with a mixture of honey and water. But he mendaciously tells the people at home that 'the honey was not fat' (Bleek 1924:2). Interestingly, //Kabbo, himself a shaman (Chapter 2; Lewis-Williams 2002a:15–17), made more of /Kaggen's collection of honey than did /Han≠kass'o in his performance. Indeed, //Kabbo seems to have been more interested in shamanic affairs than some of the other /Xam informants.

Because the Mantis is not bringing home honey, /Kwammang-a sends his son, the young Ichneumon, to spy on him. He reports back that the Mantis is giving the honey to the eland. One of Lloyd's notes shows that this news causes the Mantis's family and the Meerkats to 'conspire together' (L.II.4.503'; Bleek 1924:3). The /Xam word here is *kwaitenkwaiten* (L.II.4.503'); it may be translated as 'whisper' (Bleek 1956:112). But Lloyd's note gives a vivid impression of conspiracy and plotting: 'They conspired together, the ichneumon whispered, /kuamman-a also whispered, the miercats also whispered'. Now there is dissension in the Mantis's family.

/Kwammang-a then goes to the waterhole and imitates the Mantis's call. The eland, hearing the call, emerges and /Kwammang-a shoots it, thus setting in train the next sequence of events. The call of the Mantis is an interesting point: he called the eland out of the water. Qing reported a related idea. When Orpen asked Qing, 'Where is Cagn?' he replied: 'We don't know, but the elands do. Have you not hunted and heard his cry, when the elands suddenly start and run to his call?' (Orpen 1874:3). The Mantis can summon eland. I suspect that the call of the Mantis was a widely held concept.

While the Meerkats conspire, the Mantis is seeking honey elsewhere, but all that he finds is 'dry'. He interprets this as indicative of social discord. He suspects the Ichneumon of causing this problem by reporting to /Kwammang-a's people what he had seen at the waterhole.

Although no /Xam word translatable as 'charms' is used here, the Mantis nevertheless knows what is happening afar, as did Cagn by means of his charms (Chapter 2). Instead of charms, it is honey that suggests to the Mantis that trouble is afoot: he goes to two bees' nests and, speaking of the honey as if it were meat, finds that in both places it is unexpectedly 'dry' and 'lean', whereas it is usually 'fat'. He concludes that 'danger has come upon my home' and that 'blood is flowing' (Bleek 1924:3). The word translated 'home' is *//neiŋ*; it means an individual hut or a camp comprising a number of huts (Bleek 1956:618). However, in the eland myth, the *//neiŋ* is not the Mantis's camp but the place where the eland was killed, that is, the waterhole itself. Lloyd added an explicit explanatory note to this effect: 'where the eland lived, not /kaggen's own abode' (L.V.1.3640'). That the mediatory waterhole should be spoken of as the Mantis's personal 'home' signifies his status as one who moves between cosmological realms. It follows that the 'danger' that he fears is not the dissent at his home camp, the whispering, but rather the death of his eland at the waterhole. The /Xam word here translated as 'danger' is */a:*. Dorothea Bleek's *Bushman Dictionary* gives it as 'fight, harm, curse' (Bleek 1956:267). In the Ju/'hoan language, the word for 'fight' also means a substantial concentration of supernatural potency. As we shall see, */a:* recurs in the next chapter, where, somewhat confusingly, it has two meanings in a single sentence.

Honey and Potency

In San thought generally, the honey that the Mantis obtains by means of his shamanic powers in Qing's myths and that, in the /Xam narratives, he feeds to the eland and rubs on its flanks is rich in connotations. It is different kinds of honey that give eland, gemsbok, springbok, quagga, and hartebeest their distinctive colours (Bleek 1924:10). It features in numerous myths from different San linguistic groups and from vastly different natural environments. We need to explore its significances further.

As we have seen, honey is, for the Ju/'hoansi, anomalous in that, along with fat, it can be both eaten and drunk (Biesele 1993:86). In a further extension, the Ju/'hoan euphemisms 'to eat or drink honey or fat' mean to have sexual intercourse. Biesele (1993:100–01; 1978:927) concludes that fat and honey are 'symbolic of the great mediation between men and women—sexual intercourse'. When the Mantis feeds honey to the eland in the waterhole his action may suggest sexual creation in an oblique way. This idea was not restricted to the /Xam and to eland. Guenther found that there are 'widespread stories that depict [the trickster] fondly raising one or several baby antelopes (or, in a farm-based variant, lambs), which he feeds with honey water and other magical, strength-inducing medicines' (1999:100–01).

Biesele takes the matter further:

> Many clues, too, point to a delicate use of this same sexual metaphor [eating and drinking honey or fat and the killing and eating of herbivores by carnivores] to describe the collaborative processes of the trance dance. Men and women, and the objects and qualities in their respective spheres, are actors in the symbolic dramas of folktale and dance. (1993:87)

As Biesele found, the metaphors of fat and honey permeate San shamanic activities *and their folktales*: indeed, the power of honey finds its way into diverse areas of San life. Honeycombs and even individual bees also appear in San rock art (Figure 3.3).

Given the associations and contexts that I have so far mentioned, it is perhaps not surprising that honey and bees were (and still are) believed to have much supernatural potency. This quality is seen in various San customs. A young San man who has recently become a shaman by receiving potency from his mentor or directly from god should avoid eating powerful honey for a few years (Heinz 1966:141; Marshall 1999:55). Then, after his first dance as a fully fledged healer, a young shaman 'should go hunting and should not give up until he has killed a large animal' (Marshall 1999:54). Nevertheless, even a mature hunter who is tracking an antelope should not eat honey because its potency will neutralise the poison on his arrows (Marshall 1999:158). Again, we must note that potency is neither good nor bad: its concentration is what counts. Linking the provision of both honey and meat to social esteem, Alan Barnard makes an important point: among the Nharo San, he found that men 'have the prestige of being considered good hunters or good honey-collectors' (1992:142). Seeking honey is thus in some ways akin to hunting (Marshall

FIGURE 3.3 A tracing of a San rock painting of trance dancers and bees. A figure wearing an eared cap bleeds from the nose and bends forward as he reaches out to a strange creature that appears to hold dancing sticks.

1999:54). The importance of prestige, rather than overt wealth, should be kept in mind when discussing San egalitarianism, a point to which we shall return.

Potency is at its strongest in a trance dance (Katz 1982:94). Just how honey potency works in a trance dance is seen in a Ju/'hoan medicine song named 'Honey'. Transformation is involved in the performance of the song: 'One man speaking of the Giraffe Songs [. . .] said, "When he (a man) dances the Giraffe Dance he 'becomes giraffe'." Another said the same of the Honey dance—he 'becomes honey"' (Marshall 1999:73).

A shaman can transform into honey: the thought is arresting. Trancing, transformation, the provision of honey, and resulting prestige are all interwoven. A Ju/'hoan person may obtain the powerful Honey Song through interaction with god:

> [A man named /Gao] had received a Honey Song. It was daytime, and he was awake when //Gauwa [the lesser god] appeared and bid him follow to a tree that had a beehive in it. //Gauwa pushed /Gao into the hole in the tree where the baby bees and the honey were. Bees and honey both have *n/um* [potency]. Thus //Gauwa gave him the *n/um* of the Honey Song. //Gauwa then taught /Gao the music of the song. (Marshall 1999:76–77)

Here (presumably imaginary) physical contact with honey, potency, and its associated song come together.

As we have seen, the Ju/'hoan Great God's wife is known as Mother of the Bees. She can sometimes lead men seeking honey to a bees' nest (Marshall 1999:6, 32). ≠Gao N!a himself is very fond of honey (Marshall 1999:31). He can in fact change himself into honey when he capriciously wants to lure a man to his death (Marshall 1999:31). Similar ideas existed among the /Xam. As we have also seen, the Mantis's wife, the Dassie, was said to live with the bees (Bleek 1924:47). Honey is thus closely associated with supernatural beings and contact with them.

The way in which the obtaining of honey can lead to a dance (a boon the Mantis denies his family) is described in an account that Lloyd obtained from /Han≠kass'o (Bleek & Lloyd 1911:353–59). First, the people 'beat the *!goïn !goïn* ["bull roarer"]' to cause the bees to swarm and to move from one place to another, presumably from one bees' nest to another owned by the narrator's people: again, groups of people compete for possession of honey. Then, in an echo of what should have been happening in the creation myth, 'the men take honey to the women at home' (Bleek & Lloyd 1911:355).

There is probably a suggestion here of the sexual euphemism 'to eat honey', but, either way, the theme of bringing home the honey and its beneficial consequences beyond the satisfying of hunger is explicitly dramatised. After they have eaten, the people dance until dawn, which, the Ju/'hoansi say, is a time of especially strong potency. The same appears to have been the case for the /Xam. When the sun rises, it 'shines upon the backs of their heads'; a note by Lloyd explains that this means 'the holes above the nape of their neck' (Bleek & Lloyd 1911:357). It is from such 'holes' (*n//ao*) that Ju/'hoan shamans speak of expelling sickness.[20] /Han≠kass'o did not explicitly say this was a trance dance, as the narrative suggests it was: he took it for granted.

The status of the eland is strikingly evident in the Ju/'hoan men's 'respect word', the word they use when out hunting eland. They refer to the eland as *djxani* (formerly given as *tcheni*), their word for 'dance' (Biesele, pers. comm.). By using this word, they recognize that they are hunting not just meat but also supernatural potency that they will be able to exploit in the trance dance that will almost inevitably follow an eland kill. By virtue of its huge amount of fat, the eland became the San shamans' power animal *par excellence*: eland potency is more desired than that of any other animal (Lewis-Williams 1981).

This point brings us to what I suggest is the central meaning of the creation of the eland myth—if the myth can be said to have a 'central' meaning rather than a cluster of meanings. Ju/'hoan hunters like to dance next to the carcass of a freshly killed eland because, as they say, the place is filled with potency. When the hide of an eland is stripped back, a sweet scent arises. A nineteenth-century Western explorer wrote of an eland kill: 'As I stripped away her sleek, smooth coat, a strong sweet perfume, redolent of pleasant herbs, came to my nostrils' (Bryden 1893:422). The /Xam San interpreted this odour as deriving from honey. A link to another antelope and then back to eland is seen in a remark given by //Kabbo: 'The kudu eats honey; therefore the kudu's scent is like the eland' (L.II.3.464). Generally, the San regard scent as a vehicle for potency and the transference of potency. The killing of an eland implies contact with the supernatural by means of its great potency.

Honey, the 'scent' of honey, fat, the eland, and the trance dance thus come together in a powerful nexus. Dispute over eland/honey is thus a dispute over the possession of potency and, consequently, rivalry for access to the supernatural.

More Than Just an Antelope

Having come thus far, we may conclude that the creation of the eland, seen in the context of San thought networks, is not merely the creation of an antelope that people can hunt and eat. Rather, the narrative concerns the creation of the eland as epitomising, or embodying, the whole San belief and ritual system for contacting the supernatural. That disputes over the possession of this antelope should subsequently arise is understandable. Further supernatural implications of the eland are seen in the Mantis's subsequent fight with the Meerkats.

CHAPTER FOUR

The Fight with the Meerkats

He seeks the eland's spoor, he sees the blood.
(//Kabbo, L.II.4.507)

This chapter takes us on from the creation of the first eland to its death and the Mantis's ensuing fight with the Meerkats. (For a summary of the whole myth see Chapter 3.) Here we find the other side of the coin, as it were: creating is all very well, but maintaining possession of what one has created can be a problem. As we saw in Chapter 2, this is especially so when the disputed possession is supernatural potency (the eagle's honey), the 'essence' that underwrites so much of San life.

The narration about the fight with the Meerkats that I now discuss was given by //Kabbo. The Bleek family seems to have considered him different from /Han≠kass'o, Diä!kwain, and the others. It was his portrait (Colour Plate 3) that Wilhelm Bleek sent to the Egyptologist Richard Lepsius and that Lucy Lloyd selected for the frontispiece to *Specimens of Bushman Folklore*. In 1873, Lepsius showed it to the Berlin Society for Ethnology, Anthropology and Prehistory (Bank 2006:186).

//Kabbo's life is crucial to understanding the nature of the conflict with the Meerkats. He came from the semiarid plains near the present-day town Kenhardt (Colour Plate 1).[1] At about 60, he was older than Qing, /Han≠kass'o, and Diä!kwain. Bleek chose him early on as one of the best narrators from among twenty-eight San prisoners who were incarcerated in the Cape Town Breakwater Prison. He had been imprisoned for sheep stealing, a recourse to which he was driven by hunger: 'I do shoot, feeding myself. For, starvation was that which I was bound, on account of it, for starvation's food' (L.II.32.2924'). He arrived at the Bleek's suburban home on February 16, 1871, and remained with them until October 15, 1873, when, longing for

Myth and Meaning: San-Bushman Folklore in Global Context by J. D. Lewis-Williams, 97–116

his wife, he returned home. He hoped to return to Cape Town with her, but he died on January 25, 1876 (Lloyd 1889:1).

His time with the Bleek family was highly profitable from the point of view of their linguistic and folklore researches. Within four months of his joining the household his term of penal servitude expired, but he decided to remain with them for a while, enticed to some degree by Bleek's promise of a gun. Lloyd wasted no time and recorded her first work with him on February 23, probably after a few days of informal discussions to establish rapport. Her initial records show a hesitant beginning, but //Kabbo eventually became an excellent narrator and 'patiently watched until a sentence had been written down, before proceeding with what he was telling' (Bleek & Lloyd 1911:x). At the beginning, both //Kabbo and Lloyd herself were inexperienced in a task that, at that time, had no precedent. Nevertheless, he contributed over 3,100 pages of material. Lloyd reported that 'he much enjoyed the thought that the Bushman stories would become known by means of books' (Bleek & Lloyd 1911:x). A century later, working in the Kalahari with Megan Biesele, I found that the Ju/'hoan San were similarly anxious for outsiders to know the truth about them and to understand what they believed (Lewis-Williams 1981:35–36). Unconsciously, they were echoing what //Kabbo had said so long before, but they were consciously providing additional incentive for the writing of this book and for Biesele to continue with her multifaceted work on their behalf.

Writing years later of //Kabbo, Dorothea and her sister Edith Bleek recalled that he 'was great in story-telling'. The Bleek children delighted in watching his 'eloquent gestures'. Their favourite piece 'was the death of a monster who had frightened and was going to eat the children. The awful noise this monster made, and his final death-splutter over a hot stone, when [//Kabbo] clapped his hand to his mouth and rocked himself in agony, could not have been better done' (Bleek & Bleek 1909:42). His painstakingly recorded texts are, however, more sober affairs, shorn of all the excitement of performance.

Nevertheless, they are of particular interest because //Kabbo was a *!khwa-ka !gi:xa*, a shaman who controlled rain (Lewis-Williams 1981:27). This became apparent as a result of a curious incident. While he was with the Bleek family, he was required to work in their garden. On one occasion he found the ground hard, and so, by 'dreaming', as he put it, he caused rain to fall and soften the ground. 'I dreampt that I told the rain to fall for me. . . . The rain assented to me' (L.II.6.625).

By that time he had rightly come to consider himself not so much a servant as one who taught Bleek about San life (Bleek & Lloyd 1911:315). He became affectionately known as the 'Philosopher'. Bleek's daughters described him: 'This gentle old soul appeared lost in a dream-life of his own' (Bleek & Bleek 1909:38). Today it is customary to see statements like this as paternalistic and leading to the romanticized image of the San later publicized by such writers as Laurens van der Post. Certainly, they make us uncomfortable. However, if we set the Bleek family in the context of what were the nineteenth-century colonial views of the San, I believe they stand out as altogether more humane: indeed, they openly challenged colonial attitudes by taking an interest in people whom many regarded as virtually subhuman. In any event, Katz (1982:236) found that Ju/'hoan shamans seemed 'predisposed to kia ['trance'] by dint of their fantasy lives. . . . Since fantasy life and kia are both altered states of consciousness, experiences with one state may affect the other'.

San names are often illuminating. They reflect incidents in people's lives and characters (Lewis-Williams 2013b; McGranaghan In press). Given the manner of his rainmaking, //Kabbo's /Xam name appropriately means 'dream'. He was familiar with the spirit realm which, the /Xam believed, could be accessed by dreaming. His second /Xam name, /Uhi-ddoro ('smoking tinderbox'), probably links him to the Mantis, one of whose names was //Kandoro ('tinderbox owner'). More explicitly, //Kabbo was said to be a */kaggen-ka !kwi*—'a Mantis's man' (Bleek 1936:143–44). When he narrated tales about the Mantis, such as the one we examine in this chapter, he was on home territory. The Cape Dutch name he acquired from the white settlers is also of interest: it was 'Jantje Tooren'. Knowing that he was a rainmaker, we may perhaps speculate that the second part of this Dutch name derives from what is now the Afrikaans word 'towernaar', magician. He may have explained to colonists what '//Kabbo' meant, and, as a result, they may have bestowed on him a Dutch near-equivalent.

When he told Lloyd the story of the fight with the Meerkats, he was in the early stages of adjusting to the unfamiliar circumstances of dictating a narrative and having to wait until what he had said had been written down, phrase by phrase, sentence by sentence. A rather staccato and sometimes jumbled format was the result. At one point Lloyd wrote on a verso page: 'later put in here, but does not quite fit, in the latter part' (L.V.II.503'). Like Joseph Orpen (Chapter 2), Lloyd was expecting a more ordered form of story-telling. Even so, it would, of course, be wrong to assume that //Kabbo would have spoken like this in his home circumstances where the telling of a tale was a dramatised performance. Then we must remember that Lucy Lloyd did not note down any questions she may have put to him during the process of recording his tales, so we do not know if some of his repetitions were part of his own style of narration or whether they resulted from requests for clarification.

Bleek and Lloyd faced another major barrier to understanding what they were being told. We find this difficulty especially in the text that this chapter addresses. They initially knew next to nothing about the San beyond the colonial stereotypes of a wild and religionless people, but, as their work progressed day by day, they gradually learned unexpected things. They were repeatedly coming up against beliefs, and the idiomatic phrases that expressed those beliefs, that were entirely new to them. Narration, discussion, and translation had to proceed hand in hand. Often //Kabbo's /Xam words hold allusive meanings that Bleek and Lloyd could not have been expected to know.

To give some idea of how Lloyd managed the task of recording, I give below a complete, lightly edited transcription of her notebook pages L.II.4. 493–94, 504–13. My version is edited as to punctuation and the standardization of tenses and names. I have also added paragraph breaks. The concept of a paragraph was, of course, foreign to //Kabbo. I have included in my transcript Lloyd's notes that seem to record answers to pertinent questions that she put to him.

The narrative begins as the Mantis leaves the growing eland in the waterhole after he had fed it honey and rubbed its flanks with honey mixed with water (Chapter 3). He was accompanied by the Ichneumon. He told the Ichneumon to sleep under a kaross, but the child peeped out and saw what was happening: the Mantis was calling the eland out of the water and giving it honey. His young grandson thus became privy to his secret. A child as spy is a motif that recurs in /Xam tales.[2] What is important in this one is the relationship between /Kwammang-a's son and the Mantis. Because of the joking relationship between alternate generations, it was natural for the Mantis to take the Ichneumon with him to the waterhole. At the same time, the child was, as his name implies, from a different conceptual group: pawed creatures. Having seen what was happening at the waterhole, the Ichneumon tells his father to imitate the Mantis's actions and to call the eland. 'Calling an eland' seems to have been an important concept, as the epigraph to Chapter 2 suggests. Conflict is again to the fore. But what sort of conflict?

THE FIGHT WITH THE MEERKATS

Narrated by //Kabbo

September 1871

[493] The eland goes into the reeds. It sits down. The Mantis arises and picks up the quiver, he slings it on, he walks. He says, 'O Ichneumon, let us return home.'

They talk. He says to the Ichneumon, 'O Ichneumon, you are crazy.' The Ichneumon says, 'I am not a rascal!'

The Ichneumon goes and tells /Kwammang-a, 'O /Kwammang-a, [504] it must be that which eats the honey. It is not small. It is white. It has gone into the reeds.'

/Kwammang-a says, 'It is my shoe. You must take me, so that I may see it.'[1]

The Ichneumon said to /Kwammang-a, 'You must go and cut honey. You [505] must take the honey to the water. You must call it, while you moisten the honey. Then you will see it.'

Then /Kwammang-a did so. He took the honey to the water and moistened it. He called the eland. The Ichneumon told him that the Mantis had said, '/Kwammang-a's shoe's heel.' /Kwammang-a said that it was his shoe's heel.

[506] The eland leapt out of the reeds. It trotted up. The Ichneumon spoke as the eland trotted up. He said, 'You must wait for it.'

The eland trotted up, the eland approached. It drank water. /Kwammang-a shot it as it drank water, as it stood. It sprang back. It sprang away. Springing, it went away feeling that [507] it would go quite away. It went away to die on the ground.

Then the Mantis came and missed it. He called it. It did not come. Then the Mantis cried, for he felt that he did not see the eland. The Mantis got up. He sought the eland's spoor. He saw the blood. Again, he cried.

He picked up his kaross. He [508] covered his head. He returned, going along. He cried. He goes and lies down while the sun stands. He lies down. He is angry. Angrily, he lies down.

/Kwammang-a goes and speaks to /k'i ya //koe, a Meerkat. The other Meerkat is here. The other Meerkat is here. The other Meerkat is also here. The other Meerkat also is here. /k'i ya //koe is here. /Kwammang-a is here. They go to the dead eland. [509] /Kwammang-a looks at the dead eland. /Kwammang-a returns,[2] while /K'i ya //koe brings it.

The Mantis goes to them. Then a Meerkat goes and, wrestling, throws the Mantis on the eland's horns. Being angry, he wrestles with the Mantis and throws him on to the eland's horns,

It was because the Mantis had pierced open another eland's gall, that was why the Meerkat was angry with the Mantis. The Meerkat says the Mantis always acts like this: he pierces the eland's gall open [510]. Therefore the Meerkat is angry with him; he came to them in order to pierce open the gall which was ??? in the night. Therefore the Meerkat fights him.

Then he slowly returns home, while the Meerkats are packing up the eland meat. They put it in the kneeboom tree. The Mantis goes to lie down for his head aches. He lies when the sun is up. Then the sun [511] sets while he lies, because his head aches.

He sleeps, he trembles as he lies. Then the tree[3] comes out of the ground on which the meat is placed. The Meerkats' aprons are on it. The Meerkat's back aprons are on it. The Meerkat's quivers are on it, standing on it. The bows lie below. The Meerkat quivers bring up /k'i ya //koe's quiver. His [512] apron is placed on it. /K'iya //koe's back apron is placed on it. His quiver hangs on it. His bow lies underneath. Buttons are on it. The Meerkat's buttons are on it. The other Meerkat's buttons are on it. All their buttons are on it.[4]

While the Ichneumon comes to take the Meerkat's !kwai [prob. BD 458, hole]. The Mantis takes /K'i ya //koe's quiver. [513] The Meerkat returns home. He is not dressed. Then his wife sees him. His wife says, 'What have you done, that you have not brought the eland's flesh. That must be why you have not brought the quivers. That must be why you have only single arrows in your belts.

Lloyd's notes

1. [503'] Later put in here, but does not quite fit, in the latter part: You must take me, that I may see it. They conspired together, the ichneumon whispered. / Kwammang-a also whispered, the Meerkats also whispered.
2. [508'] takes out his arrow and returns to his home.
/K'i ya //koe: ander Bushman, firste Bushman.
3. [510'] a certain tall tree, the 'crie boom'.
The tree comes out of the ground, rises up, with all the things hanging on it, and goes then thro' the heaven and comes down near the Mantis's head where it descends near his head and grows there, making a house for him, as he lies.
4. [511'] All the bows are on it. It itself mounts up into the sky, at night, of itself it goes to stand on the ground, because the Mantis has ?conjured it.

Arrows and Rituals

The arrow that shot the eland probably had specific significance that //Kabbo would not have thought to explain to Lloyd.[3] First, among the Kalahari Ju/'hoansi, a shot animal belongs to the man who owns the fatal arrow (Marshall 1961:237–38)—in this case it would probably be /Kammang-a. Marks on arrows signify the Ju/'hoan owners, and a man may exchange arrows with an expert hunter if he feels that his own hunting is in an unsuccessful phase: a more successful hunter may thus help a man out. The practice of arrow-sharing for meat explicitly includes women (Biesele, pers. comm.). If a similar principle obtained among the /Xam, it seems that, although the Mantis created the eland, the kill belongs to /Kwammang-a.

A note on 508' suggests that a Meerkat pulled the arrow out of the eland. The name of this Meerkat is given as /K'i ya //koe, and Lloyd identifies him as 'ander Bushman, firste Bushman'. This is an early instance of the recorders' encounter with beliefs about 'people of the Early Race', who were said to have lived in a mythical time when people were animals and vice versa (Chapter 3). This is a pervasive component of /Xam folklore that informants did not always feel a need to identify, but, in response to Lloyd's question, //Kabbo explicitly situated the whole tale in the mythical epoch.

Second, we may wonder about a series of complex observances that /Xam hunters followed in order to thwart the Mantis's attempts to effect an eland's escape.[4] For instance, it was believed that men following a wounded eland's spoor should not step over it. Then, the man who shot the eland returns to camp slowly, otherwise the eland will move swiftly: there is empathy between hunter and prey, a common experience among the San. When at home, if the hunter feels an itch, he does not scratch it because he 'he knows that it is not a louse, but it is the Mantis who is trying to cheat him' (Bleek 1932:236). Did the Meerkats follow these and other observances and so try to frustrate the Mantis's attempts to save his eland? We do not know, but this may be another of the 'taken-for-granted' parts of the myth that could have been dramatised in some performances but not in others.

An element of these eland hunting observances is, however, of special interest in our present enquiry. San arrows are composite: principally, they comprise the reed shaft, a torpedo shaped link, and the bone or stone point itself. In more recent times, points were, and still are, made from iron, nowadays often fencing wire. In the Kalahari, the poison is placed just behind the point so as not to blunt it (Lewis-Williams 1981:55–56, 58–59, 66). When the arrow hits an animal, the link splits either the shaft or the small collar that joins it to the point. The shaft and the link then fall away and do not work loose as the animal runs: the poisoned point remains embedded in the animal. The hunters then search for the shaft and the link to see if any of the animal's hair adheres to it: the hair confirms the species of animal and that it has been wounded (Bleek 1932:235).

When Diä!kwain gave Lloyd his account of these observances, he added a significant observation that she recorded on two verso pages (L.V.17.5329', 5330'). He said that the link, or 'shank' as Lloyd translates *//kăbba*, was made of eland bone. On the face of it, other species of bone would do as well, and, pragmatically, the San probably did use other kinds of bone as well as eland. But, because they accept a continuity between the raw material and the finished product, it seems possible, even probable, that the eland bone link was believed to have eland potency. Qing told Orpen (1874:5) that the Mantis was in the elands' bones and caused the elands to escape hunters. The first eland itself, it will be recalled from the /Xam creation tale, was made from eland hide (the shoe), and it was killed by an arrow that probably had a piece of eland bone as a component.

Levels of Significance

So far, we can see that the narrative communicates on three levels. Each takes us successively deeper into the significances of the tale. The first derives from the /Xam hunting observances I have mentioned. Hunters (the Meerkats) have killed the Mantis's favourite animal, the eland, and he wants to punish the offenders. At this, I believe superficial, although not entirely irrelevant, level the Mantis places a high value on his 'son' (the eland), while *at this juncture* /Kwammang-a and the Meerkats seem to regard the creature simply as food to be hunted and eaten. As we have seen, the Mantis contrives the escape and survival of a wounded eland by various ruses. Now, his being thrown on the eland's horns in the fight with the Meerkats is probably an ironic inversion of one of the Mantis's ways of resuscitating a wounded eland. Diä!kwain put it like this: '[The Mantis] goes and strikes the Eland's horn, and the Eland arises, the Eland eats, because it feels that it has quite come to life by means of the Mantis's doings, although it had nearly died' (Bleek 1924:12). The word translated 'doings', *ddiddi*, is elsewhere given as 'magic', although generally its restricted denotation is prosaic 'doings'.[5] In the myth, by contrast, instead of striking the animal's horn and so causing it to arise, the Mantis is himself thrown on to it and it hurts him. His way of saving a wounded eland is perverted.

On a second level, the Mantis attacks the Meerkats because they had captured his child. In San thought and idiom, hunting and mating are symbolically equivalent (Marshall 1959:354; McCall 1970). In terms of this equivalence, /Kwammang-a's shooting of the eland and the Meerkats' appropriation of it may be taken to symbolise the forced marriage of the eland into the Meerkats' family. This 'marriage' is thus an inversion of /Kwammang-a's marriage to the Porcupine. In /Kwammang-a's marriage to the Porcupine, he, the man, is legitimately performing bride-service for the Mantis at the Mantis's camp. By contrast, /Kwammang-a's shooting of the eland violates the norms of bride-service in that 'she' (though the young eland is said to be male) goes to her husband's (the hunters') camp. The Mantis was thus resisting the surrender of his 'child' to a family of pawed creatures who were his affines. This sort of dissonance is not spelled out in the narration; it is latent—taken for granted—and contributes to the sense of unease that the Meerkats' actions would create in San listeners' minds.

The third level of understanding is the richest and probably the most essential level of the myth. It ties in with significances implied by the nuggets I discussed in Chapters 2 and 3, and indeed in other San tales as well: shamanic belief and practice. The Meerkats stole not just an animal for food but, more especially, the supernatural potency with which the Mantis had imbued the eland when he fed and anointed it with honey.

The Eland's Potency

The effect of this unauthorised appropriation of potency is described in a performance of the myth given by Diä!kwain (L.V.1.3648–52). Although comparatively short, his account of the conflict that ensued between the Mantis and the Meerkats includes nuggets that fill in much of the deeper significance of the narrative.

[3648] And the Mantis turned back, on account of it. He cried, and he took up the quiver, on account of it, he ran along the eland's spoor [3649] while he cried. And he ran going out on a little bank, and he saw the Meerkats as they were cutting-up the eland. The young Ichneumon was with them.

The Mantis slung the quiver over his head[1] because he saw [3650] the people. He saw the young Ichneumon with them. And he now said, 'These things must be those which I said them, that the young Ichneumon must have done.[2]

The people are standing here cutting-up.' He took [3651] out an arrow. For, he intended to fight the eland's fight. He ran up to the people on account of it and planted his foot. He shot at the people, and the arrow came back.[3] It passed over his head and he [3652] avoided it. The Meerkats continued standing as they cut[4] the cooked flesh.

They fed themselves with the eland's flesh which they were cooking. They knew that these were not arrows that would kill them. Therefore they continued at rest.

Lloyd's notes

1. [3648'] His arrows were now more free, and he could draw out the arrows with his right hand from the quiver which was now on his back on the right side.
2. [3649'] for people now are those who cutting up stand here.
3. [3650'] this is the custom of poor /kaggen's arrows.
4. [3651'] They were not afraid of him, because they knew that his arrows would not come to them.

In Dorothea's published version of this passage, Diä!kwain says that the Mantis intended to 'fight the Eland's battle' (Bleek 1924:4). As the preceding transcription shows, Lloyd's manuscript has 'fight the eland's fight' (L.V.1.3650). In Dorothea's wording, the sentence seems to mean that the Mantis intended to fight *on behalf of* the eland. This translation is another example of the confusion that sometimes arose from Lloyd's unfamiliarity with /Xam beliefs about potency and shamanism. Puzzled by the repetition of 'fight', she resorted to a typical Western phrase: to fight someone else's battles for them. The /Xam word that she translated as both the verb 'fight' and the noun 'battle' is */a:*. We encountered it in the previous chapter where we saw that the *Bushman Dictionary* also gives it as 'fight, harm, curse' (Bleek 1956:267). In Diä!kwain's performance of the death of the eland Lloyd gives */a:* as 'danger' (L.V.1.3640). The Mantis says: 'It seems as if danger has come upon my home' (Bleek 1924:3). This is when he finds that the honey is 'dry' or 'lean', and, as we saw, the 'home' is the waterhole where the eland was both created and killed.

The translation of */a:* is a crux that illuminates much of the narrative. As a nugget, it merits close attention.

One of the meanings of */a:* is clarified by the way in which the Ju/'hoansi use the word. Although the Ju/'hoan and /Xam languages are largely lexically distinct, both have this particular word. In addition, a 'fight' can, for the Ju/'hoansi, mean a dangerous concentration of supernatural potency.[6] Bearing in mind this and the various uses of */a:* in the /Xam language, I argue that the passage in question would be better translated to mean that the Mantis intended to fight *against the eland's great potency*, which had been appropriated by the Meerkats. At once the fight begins to make sense. When the Mantis shot at the Meerkats, his arrows were deflected back at him, not just because he is a trickster but because of the potency they had 'stolen' from him. The eland's potency, now in the hands of the Meerkats, was too strong for him. He had to dodge the arrows.

This interpretation is borne out by the observation that, in the Kalahari, the Ju/'hoansi like to dance next to the carcass of a freshly killed eland. They say, the place is redolent with the antelope's scent and released potency. In recounting a myth, a Ju/'hoan speaker said: 'When they had cooked the meat for them to eat, he said, "Now serve the meat so that we can get a dance started. My children have killed an eland, so let's praise it by dancing"' (Biesele 1996:153). The shamans are then able to harness the potency for a particularly effective trance dance. Dancing next to an eland kill, they enter trance and cure all present of ills known and unknown. The hunters do this even if there are no women there to provide the clapping accompaniment because they have tracked the eland far from their camp.

During such a dance, the spirits of the dead, attracted by the beautiful dancing and singing, shoot small, invisible arrows-of-sickness at people. The dancing shamans, fortified by the eland's potency, deflect these arrows and remove any that may have penetrated people. On these occasions, one could say, the /Xam mythical event is still being acted out in the Kalahari. I therefore think that the image presented by the Mantis's fight with the Meerkats next to the dead eland is, in essence, that of a San curing dance and thus an outcome of the way in which the Mantis imbued the growing eland with potency in the first place.

Still on the shamanic level, we can consider the curious role of the eland's gall. It is not clear why the text reads 'another eland's gall'; perhaps what happened in the myth applied to any eland kill. Either way, in //Kabbo's performance, the Mantis angers the Meerkats by piercing the eland's gall, an act that causes darkness. Other performances develop this episode at greater length. In them, the Mantis first leaps into the darkness created by the gall and later creates the moon. /Han≠kass'o gave the incident like this: 'Then he pierced the gall, he made the gall burst . . . , and the gall broke covering his head; his eyes became big, he could not see. And he groped about feeling his way' (Bleek 1924:9). Then, as we saw in the previous chapter, the Mantis wipes the gall from his eyes with an ostrich feather, which becomes the moon.

Clearly, the gall is a highly significant nugget; it must be related in some way to the conflict between the Mantis and the Meerkats. Other references in the Bleek and Lloyd Archive help us to understand what is transpiring. When Diä!kwain was telling Lloyd about what happens to people after death, he said: 'And our gall, when we die, sits in the sky; it sits green in the sky, when we are dead' (Bleek & Lloyd 1911:399). The word Lloyd translated as 'green', */kai:n*, may be translated 'green, yellow, shining'; as a verb, it can mean 'to light' (Bleek 1956:297). In this instance, it seems that gall is associated with the spirit, or essence, of a person. It is also significant that the importance of an animal's gall is emphasised by Ju/'hoan shamans, who sometimes eat the gall of a lion in the belief that it is the centre of that animal's power (Wilmsen, pers. comm.). Particularly powerful shamans are believed to have the ability to walk abroad at night in the form of a lion.

Taken together, these beliefs and practices suggest that, when the Mantis pierced the eland's gall, he was releasing a deep reservoir of potency that the

Meerkats had, for whatever reason, not managed to access. It seems clear that, when he leaps into the resulting black cloud, he is immersing himself in that potency. The way in which he gropes about in the darkness graphically suggests a man disoriented by trance: 'He may stagger around and lurch into the fire, trample on the women, fall headlong into their circle, somersault over them, or crash full-length on to the ground' (Marshall 1969:376). Katz (1982) describes a man in trance finding his way 'as if groping in the dark'. In sum, it seems that the fight with the Meerkats is an instance of shamanic fighting in the spirit realm.

Another example of supernatural conflict broadens our understanding. It is the tale '≠Kágára and !Hãũnu, who fought each other with lightning' (Biesele 1996; Bleek & Lloyd 1911:113–19;). These two men were shamans, !Hãũnu being specifically a rainmaker. They fought over possession of ≠Kágára's young sister, whom !Hãũnu had abducted and married. Here there are echoes of /Kwammang-a's 'abduction' of the eland. In terms of San custom, it was !Hãũnu who should have performed bride service at ≠Kágára's camp. When ≠Kágára went to fetch her from !Hãũnu's camp, she was carrying a heavy load of belongings. At this point we find a significant nugget. During the ensuing fight, !Hãũnu 'sneezed . . . blood poured out of his nostrils'. Nasal bleeding is a clear reference to trance (Lewis-Williams 1981:78ff; Lewis-Williams & Pearce 2004a). !Hãũnu retaliated by directing lightning at ≠Kágára. But ≠Kágára 'fended him quickly off' and responded by hurling lightning back at him. Finally, when ≠Kágára resorted to 'black lightning', !Hãũnu was vanquished and 'whisked . . . up' and carried 'to a little distance'. At this time, 'the clouds were thick'. The young girl and her belongings were restored to her home.

The parallels between this tale and the Mantis's fight with the Meerkats over possession of the eland are certainly striking. In one, arrows are fired off; in the other, lightning is the weapon. Lightning is akin to arrows in San thought. If angered, the rain (*!khwa*) could thunder and 'shoot' at people (Bleek 1933:297–98). Among the Ju/'hoansi lightning is more directly linked to shamanic experience: the women's more usual word for trance is *tara*, not *!aia*. *Tara* means 'the action of lightning' (Katz 1982:165). In a Ju/'hoan tale that has echoes of the /Xam Mantis-eland-Meerkats tale the trickster G!ara 'summons lightning to strike the lions dead' because they had killed his sons (Biesele 1993:109). All in all, in the fight with the Meerkats and in the fight between !Hãũnu and ≠Kágára shamanic powers are invoked to right a wrong committed by affines. We are in a thicket of interrelated significances.

The Mantis Resurgent

We now move on to see how the Mantis, after being humiliated by the Meerkats, responded. We are told that he returned home and lay down because his head ached. According to //Kabbo, the tree on which the eland meat and the Meerkats' belongings had been placed then rose up, passed through the heavens and settled next to him at his home. Probably because she did not understand San shamanic trancing, Dorothea Bleek omitted the following sentences from her published transcription of the tale:

> [H]e sleeps, he trembles [*!khauken*], as he lies. Then the tree on which the meat is placed comes out of the ground. [510'] The tree comes out of the ground, rises up, with all the things hanging on it, and goes then thro' the heaven and comes down near the /kaggen's head where it descends near his head and grows there, making a house for him, as he lies. (L.II.4.511, 510')

Another nugget was passed over. To understand these important 'magical' events we must again examine //Kabbo's use of certain words. Here, the tale is coming to its climax.

//Kabbo said that the Mantis 'trembled' as he lay down. The /Xam word translated 'trembled' is a nugget that situates the whole story of his fight with the Meerkats. It is one of those 'elements' that Biesele (1996:145) points out may be 'merely mentioned in passing' but that have 'enormous ramifications'.

The /Xam word is *!khauken* (also *!kauken* and *!kouken*). //Kabbo seems to have taken it for granted that Lloyd would know what it implied. The *Bushman Dictionary* gives it as 'to tremble' and 'to beat', but one example that it cites points to a specific context in which the word was used: 'When he returning comes in from the place to which he had gone on a magic expedition, he trembles' (Bleek 1956:425, 445). The 'magic expedition' (*/xãũ*) is the out-of-body travel performed by San shamans in dreams and during trance dances. In other /Xam texts, *!khauken* similarly refers to the violent trembling of trance. For instance: 'The others hold him down and rub his back with fat, as he beats (*!khauken*)' (Bleek 1935a:2). In another instance, 'He beats (*!khauken-i*) when he is snoring [curing] a person with his nose' (*ibid.*).[7] Today, Ju/'hoansi shamans still tremble violently as they enter trance.[8] Richard Katz, for example, noted of a trancing Ju/'hoan shaman: 'He starts to tremble, his legs quivering. . . . His look is glazed, and his body trembles spasmodically' (Katz 1982:65). Another Kalahari shaman told Bradford Keeney (2003:46): 'The legs start to twitch and feel jumpy, which starts you dancing. Soon your thigh muscles start to tremble and the dance really gets inside your body.' Here the speaker related trembling directly to dancing. Another said: 'Sometimes I lie on the ground and start to shiver. . . . These vibrations help you to see spirits' (Keeney 1999:33). Here, the last sentence is especially significant.

Lloyd encountered another problem. Her confusion of 'trembling in trance' with 'beating' in the sense of 'striking' is explicit in her notes on the Mantis's fight with the Cat (L.II.22.1965–2042). In this tale, //Kabbo used *!khauken* to mean 'strike'—the Cat hit the Mantis. The second time he used the word Lloyd was puzzled; she was encountering an aspect of San life that was foreign to her. Her transliteration of the passage reads: 'Therefore, thou didst get feathers, as thou beating stood . . . thou ascendest the sky' (L.II.22.1986). Here, the 'beating' led to shamanic flight. Lloyd noted her perplexity: 'Can this be a passive (beaten)? JT [Jantje Tooren, //Kabbo] explains that the mantis was beat<u>ing</u>. But I am not very sure, if I have the explanation rightly' (L.II.22.1987'; Lloyd's emphasis). So the Mantis's 'trembling' when he lay down at home is another instance of how Lloyd's unfamiliarity with San shamanism, its physical effects and, especially, idioms sometimes resulted in uncertainty about exactly what the informants were saying to her.

There is a final piece of evidence as to the significance of the Mantis's trembling. Speaking of the tree, //Kabbo explained: 'All the bows are on it. It itself mounts up into the sky, at night, of itself it goes to stand on the ground, because the Mantis has ?conjured [*!khau-wa*] it' (511'). The San word here seems to be cognate with *!khauken*. It was through trembling, entering an altered state of consciousness and employing his powers as a shaman that the Mantis was able to cause 'magical' things to happen. Lloyd's narrator was therefore right: in the story of the Mantis and the Cat, it was, as he insisted, through 'beat*ing*', or 'trembl*ing*' (that is, entering trance), that the Mantis was able to grow feathers and escape from the Cat by flying through the sky. We encounter a similar use of *!khauken* in the next chapter, where I discuss 'A visit to the Lion's House'.

We can now see what the /Xam narrator meant when he said: 'He sleeps, he trembles [*!khauken*], as he lies'. The Mantis entered the spirit realm via sleep, dreaming and trembling. This happens at the end of the tale, but it casts light on all that went before. In Biesele's phrase, an easily missed element, or nugget, has 'enormous ramifications'.

Buttons

Among those ramifications there are other intriguing points. One of them opens up a possible historical dimension. //Kabbo emphasises that the Meerkats' 'buttons' are on the tree along with their hunting equipment and clothes. To what was he referring?

In another context, he used the /Xam word translated 'buttons', *kunno* or *gonno*, to mean buttons on a Victorian jacket (L.II.12.1171).[9] It was probably this sort of context that led Lloyd to give *kunno* as 'buttons' when she translated 'The Fight with the Meerkats', a mythical context in which Victorian buttons could not be expected to appear. It is hard to question her translation, because //Kabbo would probably have pointed to buttons on his own Victorian garments, such as those shown in his portrait (Colour Plate 3). Lloyd had bought them especially for him, and he considered them 'handsome' (L.II. 12. facing 1171). If so, at exactly what was he pointing?

The traditional San garments mentioned in the text, 'aprons' and 'back aprons', did not require buttons in the usual Western sense of the word. So what could *kunno* mean? One answer is that the late nineteenth-century /Xam sewed buttons of Western origin onto traditional garments as decorations, as they did ostrich eggshell and trade beads. However, elsewhere in the Bleek and Lloyd Archive we encounter another context that takes us beyond mere decoration. Buttons (*kunno*) are said to sparkle and shine (*bbaiten*; L.V.12.4945–48). Sparkling was important to the /Xam because it was believed to keep lions at bay (L.V.12.4946–48). At this point Lloyd noted down two /Xam words: '*!gaiï-ten* to reflect the light (as a piece of glass $w^{\underline{h}}$ is quiet) [and] *bbă-bbaiten* to sparkle' (L.V.12.4945').[10] *Bbaiten* was also used to signify the flashing of lightning. Diä!kwain went on to explain the value of sparkling: 'The lion is a thing which is afraid of a thing which reflects the light.... the lion is a thing which does not a little fear the sparkling of a button. The buttons which she was wearing... she holding/grasping waved it about; while she desired that the button "might/should" sparkle' (L.V.12.4945–46).

In this passage, Diä!kwain spoke of 'the buttons which she was wearing' (4947), but also as if the buttons were separate objects that could be waved about. We must therefore allow that //Kabbo may have been referring principally to this *quality* of the button rather than to the fastening object itself when, as I surmise, he pointed to buttons on his own garments.

In the myth, the Meerkats' buttons are spoken of as a group of objects that seem to be valuable in themselves, not just appendages to clothes. In fact the text implies that the buttons are separate from the garments. It therefore seems likely that the /Xam word may have been applied to other shiny objects like quartz crystals, which the /Xam associated with the rain, and, possibly, its potency (Lewis-Williams & Challis 2011:130–31). Or, perhaps, white ostrich eggshell beads that they sewed onto garments and bags (Colour Plate 4). But, whatever the Meerkats' buttons were, Western objects or traditional items, it seems that they were shiny. When the Mantis took away the Meerkats' 'buttons', he left his aggressive affines vulnerable to lions, who were afraid of 'a thing which reflects the light'.

A final possibility remains. If we bear in mind the frontier situation of the /Xam, we must allow that the Meerkats' buttons may indeed have been shiny buttons that they had obtained from the colonists, even if, as objects, they were not serving the colonial function of buttons. If the shiny buttons were colonial in origin, we may wonder if the /Xam believed them to possess, beyond protection from lions, some of the power of the encroaching colonists. Does the Meerkats' possession of buttons align them in some way with the colonists and so introduce a historical element into the tale? Is an overlay of colonial tensions being superimposed on the more traditional tensions between affines? We often wonder how and in what form traditional /Xam myths survived, perhaps transmuted, into colonial times. We may have a clue in this tale.

In sum, we may say that the Mantis's taking of the Meerkats' equipment, clothing and buttons left them reverting from their human to their animal status, vulnerable and potentially starving. Whatever the answer, the Mantis's victory was decisive.

The Creation of the Moon

In addition to the creation of the eland, another creative act features in the tale. It comes at the end of the story, not at the beginning. The creation of the moon seems, at first glance, to be an aetiological tag loosely tacked on to the tale about the Mantis, the honey, the eland, and the fight with the Meerkats. The moon is, however, prominent in Khoisan belief and mythology.[11] But whether the San can be said to 'worship' the moon is another matter. Wilhelm Bleek's belief that the San worshipped 'moon, sun, and stars' (Bleek & Lloyd 1911:435) was probably formulated under the influence of Max Müller's sidereal and aetiological theory of myth that was popular at the time.[12] The idea did not survive later research by his own family.

As in dealing with other interpretative problems, we need to investigate a range of contexts in which the moon appears. We then see that the creation of the moon is not incidental to the tale, an aetiological afterthought: it is conceptually linked to the main body of the myth.

In the first place, //Kabbo explained that the moon was a shoe because 'he walks in the night' (Bleek & Lloyd 1911:53). This seems to be a non sequitur, but 'walking in the night' is probably a prudent San elision for spirits who walk at night. We shall encounter more of this sort of San avoidance custom in Chapter 9, where I discuss the 'people' who broke the string mentioned in the Prologue.

//Kabbo also said that 'another name for the moon', probably another respect or avoidance word, was '/Kaggen ka !kúken, the Devil's veldschoon' (Cape Dutch for a leather shoe; L.II.4.480'). The original owner of the shoe, /Kaggen or /Kwammang-a, does not seem to have been crucial. Its essence was what mattered: as we have more than once seen, the San linguistically elide an artefact and the raw material out of which it is made. There is another variant in /Han≠kasso's performance of the narrative. In it, the Mantis makes the moon not out of a shoe but out of an ostrich feather that had on it some of the eland's gall (Bleek 1924:9). The gall is, as we have seen, thought to be a seat of potency.

Further complexity takes us beyond the straightforward creation of a heavenly body that is useful for various pragmatic reasons, such as travelling between camps. The moon provides nocturnal light and ambiance for dancing.[13] Marshall (1962:248) found that the Ju/'hoansi 'almost invariably' danced 'at least once during the full moon' (see also Lee 1967:33). Among the Ju/'hoansi the association between the moon and the trance dance is strong.

It may be taken further. An important component of the significance of the moon is signalled at the end of /Han≠kasso's performance of the myth: 'The moon falls away and returns to life' (Bleek 1924:9). A clue to the significance of this statement came from four young Ju/'hoan boys who joined the Bleek household in 1879 and 1880 (Deacon 1996:40–43). From them Lloyd learned that the !Kung trickster's name (or one of his names) was /Xué. He was said to be like the moon because 'he dies, he vanishes, he gets up and is another person'. The boy added that, like the moon, people do not die outright (Bleek 1935b:270, 271).

Then he went further and said that, although different from the moon, /Xué could himself *become* the moon (Bleek 1935b:279). This belief is consonant with other Ju/'hoan beliefs. They still say that the ring that sometimes appears around the moon at night is the dance circle made by the feet of the spirits of the dead and that the moon is the spirits' fire (Marshall 1969:129). God's house is in the sky, the upper cosmological realm.

The waxing and waning of the moon may thus imply, probably among other things, death and resurrection and, by extension, 'death' and being 'raised again' in the trance dance, in which a deeply altered state of consciousness is spoken of as 'death'. This implication is evident in a statement given by Qing: 'Cagn gave us the *song* [power] of this dance, and told us to dance it, and people would die from it, and he would give charms to raise them again' (Orpen 1874:10; emphasis added; cf. Katz 1982:99–100). Like the trickster, shamans have the ability to return to life having 'died' in the dance. In short, they can move between cosmological realms. At full moon there is not only much light; the moon itself is fully 'raised again', as the dancers will be. The rich nexus of concepts that surround the dance may thus have included the waxing and waning of the moon, one of the Mantis's creations. Although we have no explicit evidence from the /Xam, this nexus may, for them, have included the trickster himself. In Chapter 7 I discuss parallels between trickster and trancers.

We should not expect to find these sorts of associations precisely and rigidly spelled out in San thought. Like so much else in San belief, ambiguity and, for us rather vague, associations are of the essence. If we place Western linearity and bounded meanings to one side, we begin to sense the semantic density of San myths. We may not fully understand every facet of every nugget, but we can at least glimpse the richness of the networks of meaning that the performance of San myths must have triggered.

The Mantis and the Meerkats: Conflict Resolution

The social disruption caused right at the beginning of the whole narrative sequence when the Mantis caused tensions in his family by giving the honey to the eland instead of taking it home was finally resolved when the eland meat, and presumably its potency, returned to the Mantis's camp. The Meerkats, his son-in-law's people, were defeated by the Mantis's 'trembling'.

Tension between affines is inherent in the very form of San society. All that can be hoped is that such tensions be mitigated. The mitigating mechanism posited in the myth about the Meerkats is shamanic intervention. This interpretation of the myth is born out by the Kalahari San, who still recognise that trance dances dissipate tensions. Indeed, they contrive to have men between whom there is animosity dance one behind the other, so that rhythmic unity and trance experience can reestablish emotional amity (Biesele pers. comm.). Conflict in the spirit realm merges with conflict in society: the myth is ultimately about both.

Finally, we come back to a key question about any myth. Who benefits from performances of it? Recognising the importance of shamanic beliefs and rituals for the San, a point on which numerous anthropologists have commented, I argue that performances of 'The Mantis Makes an Eland' and its sequel 'The Fight with the Meerkats' reproduced people's acceptance of the key role of shamans in coping with social tensions. San shamanism was not solely concerned with whimsical 'magical' happenings. It had a practical, social function as well, not only in the functionalist sense of maintaining social cohesion and dissipating tension but also in simultaneously underwriting social distinctions. //Kabbo, himself a shaman, set up and elaborated the denouement of his performance so that the importance of shamans was foregrounded. It was shamans who benefitted.

It is now clear that this and other San myths were not immutable givens learned by rote. Nor was the fund of episodes on which narrators could draw simply a source of amusement and entertainment, although it was that too. Rather, the whole field of myth was a complex resource that individuals could manipulate in the negotiation of social statuses (cf. Giddens 1984).

CHAPTER FIVE

A Visit to the Lion's House

Mantis jumped out of the bag,
while he got feathers, he flew up into the sky.
(//Kabbo, L.II.4.528–29)

When //Kabbo reached the end of his story about the Mantis and the Meerkats, he did not stop. Instead, he went on with three further tales. After some untranslated pages (L.II.515–19), Lucy Lloyd drew a line and then noted down his continuing narrative. This new section began with a tale that Dorothea Bleek eventually edited and published as 'A Visit to the Lion's House' (Bleek 1924:15–18). The word *//neiŋ*, translated here as 'house', is better understood as 'camp', a transitory place where there are a number of the domed grass shelters that the San construct, each with its own small fire. The *//neiŋ* is the heart of San life, and the tale takes us into the centre of San social life with all its trials and tensions.

It is remarkably structured and never seems to ramble, as do some of //Kabbo's other narrations. An exception is perhaps the Ichneumon's retelling of the doings at the Lion's house within the same narrative (L.II.5.536–46). Dorothea omitted it, but, as we shall see, it contains an important nugget. The retelling of part of a story by one of the characters seems to have been a feature of San myth performance. At the end of the tale, //Kabbo went straight into the story that Dorothea published as 'The Mantis and the Cat' (Bleek 1924:19–21). He followed that with 'The Mantis and the Tortoise', which remains only partially translated and unpublished. Although they flow into one another, it seems highly likely that each tale, and perhaps parts thereof, could be performed separately. The flow of this series of stories suggests that //Kabbo was warming to his task of narration and that he found Lloyd a sympathetic listener. The world of the San was opening up for the Bleek family.

Myth and Meaning: San-Bushman Folklore in Global Context by J. D. Lewis-Williams, 117–131

//Kabbo's full narrative about the Mantis's visit to the Lion's house can be found in Lloyd's notebooks (L.II.4.519–29, 5.530–46) and, in edited form, in Dorothea's published version (Bleek 1924:15–18). In this tale, the Mantis as a deceiving and mischievous trickster is a prominent theme, and nuggets that we have already encountered, such as the waterhole, reappear. Social themes, such as sharing and tension between affines, are also prominent. San daily affairs spring vividly to life as news of the Lions' kill reaches the Mantis's camp. Daily life is, nevertheless, seen to be shot through with supernatural elements. Owing to the length of the tale, I give only a summary.

'A VISIT TO THE LION'S HOUSE'

/Kwammang-a, the Mantis's son-in-law, told the Mantis's family that the Lions, relatives of his, had killed a quagga, and that he and his son, the Ichneumon, would go to the Lions' camp. The Mantis wanted to accompany them, but the Ichneumon warned him that he would be afraid of the Lions. The Mantis, however, got his way, and they set off.

When they came across the Lions' spoor, the Ichneumon identified the individuals to whom they belonged. Then they saw quagga's blood on the ground, and /Kwammang-a exclaimed that they would surely soon be eating quagga meat. They continued following the Lions' spoor until they came to a waterhole where they found the Lions.

The Mantis was at once afraid and asked the Ichneumon to hide him in his bag so that the Lions could not see him. He also asked the Ichneumon to place an ostrich eggshell of water in the bag so that he would be able to drink, to leave the mouth of the bag open so that he would be able to see what was happening, and to cover him with a kaross.

Soon the little Lion spotted the Mantis looking out of the bag and went crying to his mother. He said that the Mantis was whispering in the bag. When he saw the Mantis a second time, he ran to his mother again, and begged her to take what he now saw as a little hare out of the bag. When the little Lion went to the bag a third time, the Mantis winked at him and said that he would like to poke out his eye. The little Lion again ran to his mother.

The Lioness, angered by what the Mantis was doing to her child, came up to the bag and tried to stamp on the Mantis to crush him. But the Mantis jumped out of the bag, got feathers and flew up into the sky. The Ichneumon threw him up into the sky. As he went, he called upon his bag, shoes, quiver, kaross and cap to follow him. Boasting that no one was his equal, he dived into the waterhole. Coming out of the water, he called on his possessions to wait for him until he had dried himself. His possessions wanted to leave him, but he commanded them to remain there. After the feathers had been washed off and he had dried himself, he returned to his own camp.

There he deceived his wife, the Dassie, and his adopted daughter, the Porcupine, by telling them that the Lions had eaten /Kwammang-a and the Ichneumon. There were mutual recriminations as a result of the Mantis's lying. But when they saw the two coming laden with quagga meat, they realised that the Mantis had deceived them. The Mantis said, 'I wanted /Kwammang-a to come carrying quagga meat'. The Ichneumon and /Kwammang-a unloaded the quagga meat the Lions had given them. While /Kwammang-a sat in silent anger, the Ichneumon lectured the Mantis on good behaviour.

The Initial Episode: Social Normality

At the beginning of this myth, everything in the Mantis's camp appears normal and well ordered. Motivation for the tale is introduced when /Kwammang-a announces that they will go to the Lion's house to obtain a share of quagga meat. This part of the tale parallels the Mantis's request to the eagle for some honey. As in the Kalahari today, news of a major kill travels fast across the desert and visitors soon appear: a substantial kill activates social relations with all their positive and negative possibilities.

Indeed, the scene depicted in the tale is San life as it is still played out in parts of the Kalahari and, presumably, as it has been for thousands of years. Writing of the years 1803 to 1806, the early colonial traveller Heinrich Lichtenstein (1930[2]:63) described the arrival of people at a San camp after a large kill: the visitors came 'without any ceremony, or waiting for an invitation to partake of it'. In the Kalahari today, when meat is distributed on such an occasion, it is not in a chaotic scramble but along orderly lines of kinship.

The owner of the fatal arrow (as we saw in the previous chapter, he may not be the one who shot the animal) has the responsibility to gauge the quantity of meat to give to a person because that recipient will have to distribute it further to his or her relatives (Marshall 1961). Sometimes there are differences of opinion about the quantity of meat handed on, and arguments ensue.

In the /Xam myth, it seems likely that the Mantis's claim to a portion of the quagga meat was through /Kwammang-a. The 'owner' of the meat, one of the Lions, would have handed a portion to his relative, /Kwammang-a, who would have passed some of it on to his father-in-law, probably via his son, the Ichneumon. But the Mantis's fear of the Lions and his subsequent behaviour suggest that the relationship between him and his leonine relatives was strained, as it often is between affines in San communities.

Nevertheless, /Kwammang-a, the Ichneumon, and the Mantis set out for the Lions' camp confident that they will be given a share of the meat. Although they respect them, it is common in real life for people to drive lions away from their kill and claim the meat for themselves (Marshall Thomas 2006:148ff).

On the way to the Lions' camp, no tension characterises the Ichneumon's response to seeing the Lions' footprints in the sand: he and his father are evidently eager to join their kin. Indeed, the Ichneumon's recognition of the tracks left by individual members of his father's family suggests that he knows them well. The Kalahari Ju/'hoansi are still able to recognise the footprints of individuals. Lorna Marshall (1960:336; see also Heinz 1966) found that 'a person's footprints are as well known as his face'.

The Central Episode: New Dimensions

All the events so far have been 'normal' in the sense that they could have happened in any San camp. By contrast, the central episode comprises a set of 'non-normal', preternatural events, and the social order begins to fall apart. The two existential dimensions—material and supernatural—intersect with strange results. Writing about myth, the anthropologist Morris Freilich (1975:211) aptly remarked, 'Other things being equal, the "madder" the nonsense, the more hidden meanings it contains'.[1] As in previous chapters, we need to explore our records of San beliefs and religious practices to unravel these knotty happenings. For the /Xam, the 'mad' events were packed with meaning. They constitute the pivotal point of the tale.

Transformations

The non-sense starts when the Mantis gets into the Ichneumon's bag. This is a nugget (cf. Chapter 7). Bags are, of course, made of animal skin, usually that of a small antelope. As I have pointed out, the San do not feel that the process of manufacture destroys the essential identity of the material used (Silberbauer 1981:132). This is why, in mythical contexts, quivers made from springbok skin can revert to actual springbok. /Han≠kass'o gave a lively description of this mystical process: 'The skins of which people have made quivers turn into springbok, as the quivers stand about there, they get ears' (Bleek 1933:300). The artefact was never far removed from the raw material. Getting into a bag was thus in some ways like 'getting into an animal' and, I now suggest, like entering the spirit world via trance. It is the activities and experiences of shamans as they are well recorded in the San sources, not some vague sort of 'magic' (Hewitt 1986:182, 189), that explain the otherwise bizarre central section of the myth.

I begin by recalling that scent was believed to be a vehicle for the transference of potency. For instance, it was the scent of a /Xam shaman's nasal blood that was believed to keep sickness away from a person on whom he rubbed it. Diä!kwain put it like this: 'Magic things [*/kó:ö-de*] will not come where they smell the scent of another sorcerer's blood' (Bleek 1935a:35). Here */kó:ö-de* means a dangerous concentration of potency; in Lloyd's translation, 'magic things' is inadequate. Diä!kwain implies that the danger emanates from 'another sorcerer', so we again encounter the recurring theme of supernatural conflict between shamans, dead or alive. We see in it a reflection of social conflict.

Similarly, the potency of a Ju/'hoan shaman is said to be in his sweat and in the smell that comes from his burning hair: 'When the medicine men set fire to their hair [by plunging their heads into the dance fire and by rubbing coals on their hair] the *n/um* [*n/om*, "potency"] in the hair comes out in the smell, and the people breathe it in' (Marshal 1969:371). This 'breathing in' is part of the healing ritual. In other contexts, the Ju/'hoansi believe that it is the scent of a whirlwind (aptly called a 'dust devil' by some Westerners) that goes into people and harms them (Marshall 1962:239). They also believe the potency of medicine plants and supernatural substances to be in their smell (Marshall 1969:360, 371).

All these beliefs, taken together, suggest that the act of getting into a bag implied not just 'getting into an animal skin', powerful as that thought may be, but immersing oneself in the scent and potency of an animal: inside a bag, the scent of the animal from which the bag was made is strong. It therefore seems probable that the Mantis was not merely playing a prank to frighten the little Lion. He was absorbing—in Marshall's phrase, 'breathing in'—an animal's potency, the first stage of effecting his entry into the spirit world.

The concept of transformation that this action implies is further developed by the Ichneumon's next action. In response to the Mantis's request, he places a kaross over him. Karosses, like bags, were made from antelope skin, and the /Xam believed that they too could revert to antelope under certain supernatural circumstances. An 'angry' rain, for instance, was said to turn karosses back into springbok: 'Meanwhile the karosses become springbok which lie down and roll, thereby shaking out (water from their skins)' (Bleek 1933:300; original parenthesis). More relevantly, karosses were explicitly associated with shamans' activities: Diä!kwain said, 'A man who is a sorcerer will not lay down his kaross, even if it is hot, because he knows that the place will not seem hot to him, for his inside is cold' (Bleek 1935a:13). /Xam shamans preferred to be covered by a kaross when they were drawing sickness from an ailing person.

In a rather convoluted way, a certain /Xam shaman's name derived from this association between shamans and karosses. This instance is worth examining because it gives us a glimpse into the ways in which the San thought. At the same time, it shows how complex—for researchers, often opaque—San words and the tales in which they are embedded can be. Lloyd recorded a shaman's name as 'Ssũ-!kúï-ten-ttā (L.V.9.4651'–52) and translated it as 'Snore-white-lying' (Lewis-Williams 2013b:81). The significance of this name seems obscure, but Diä!kwain explained that the man had been given it by his mother because she was accustomed to heal ('snore') people without wearing her kaross, as /Xam shamans usually did. Moreover, her skin was very light in colour—'white'. /Xam shamans used to lie next to the people they were curing. They then noisily 'sniffed' sickness out of them. After that, they were said to 'sneeze' out the 'harm's things', which, they claimed, resembled arrows, little sticks or various creatures. As a shaman did this, he was liable to suffer a nasal haemorrhage (Lewis-Williams 1981:78). This healing practice was known to the /Xam as *sũ*. Probably following the early missionary John Campbell (1815:316), who described the custom and used the English word, Lloyd translated *sū* as 'snoring'. Curiously, people were said to have 'blamed' the woman because of her unusual way of 'snoring'. As a result, she named her son 'Snore-white-lying'. Embedded in this nexus of seeming *non sequiturs* is the belief that karosses were closely associated with shamanic activities.

These beliefs about karosses are expressed in rock paintings that depict shamans partially transformed into antelope: the shoulders and head are of an antelope, but lower down the human body there is a clear edge of a kaross with, sometimes, decorative animal tail tassels (Figure 5.1).[2] Often, these therianthropic images bleed from the nose, as shamans did. Both the paintings and the ethnography thus suggest that putting on, or being covered by, a kaross are, under certain circumstances, akin to taking on not only the potency but also the form of an antelope. Remarkably, the paintings help us to see what the San believed and saw in their trance experiences.

While the Mantis was in the bag and covered by a kaross, he frightened the little Lion in three ways. First, he was 'whispering' in the bag: *kweiten kweiten*. As we saw in Chapter 3, this word can mean 'to whisper, speak softly' (Bleek 1956:112), but it also seems to be associated with conspiratorial plotting. It was probably this seditious connotation of *kweiten kweiten* that frightened the little Lion. Social discord was brewing, as we saw it was among the Meerkats before they killed the Mantis's newly created eland.

FIGURE 5.1 A tracing of a San rock painting of human figures blending with antelope skin karosses (Eziko Museum, Cape Town). They bleed from the nose and are associated with an eland in a dying posture. A 'thread of light' joins a bird to the eland.

Next, the Mantis causes alarm by turning himself into a hare. The little Lion springs back and goes crying to his mother: 'O mammy, take out for me the little hare, which is in the Ichneumon's bag' (Bleek 1924:16). Why should a hare, a seemingly innocuous animal, frighten the little Lion? This nugget is explained by the specific associations of hares in /Xam myths and beliefs. The Mantis was believed to possess certain hares which distracted hunters who were after gemsbok, an animal of which he was fond—though not as fond as he was of the eland. He even turned himself into a hare: 'And the Gemsbok recovers, if we kill the hare, because it feels that the Mantis is the one whom we kill' (Bleek 1924:12). The hare was thus one of the Mantis's 'antisocial' manifestations, a creature that he capriciously used in order to thwart hunting. The little Lion had reason to be worried.

Lastly, in a particularly enigmatic nugget, 'the Mantis shut one eye and winked [*dabba*] at him' (Bleek 1924:16). Here, in Freilich's phrase, is an important 'non-sense' element. As with other nuggets, we need to follow up further San uses of the word and the concept it signifies. In the first place, more information is to be found in the Ichneumon's recounting of the incident, the passage that Dorothea Bleek omitted from her published version. We are told that the little Lion 'thought the Mantis was bewitching [*tsweiten tsweiten*, a word not in the *Bushman Dictionary*] when he did this' (L.II.5.539). Clearly, 'wink', with the arch connotations that it has in English, is not an appropriate translation of //Kabbo's word *dabba*. At this point the *Dictionary* is more helpful. There we find *dabba* translated as 'to wink, blink, twinkle, open, and shut eyes'. Only one example of its use as a noun is given, but it is highly significant: 'a person who seems to be dying, his eyes' blinking is there' (Bleek 1956:20). In the manuscript we find a fuller translation of these words: 'It was the eye blinking [*dabba dabba*] of a person who seemed to be dying. Because of that you are going to see rain water which will pour down like this' (L.V.15.5124; Bleek 1932:327–38).

Dabba is thus associated with dying, causing rain to fall (or foretelling it), 'bewitching' and, I now argue, trance experience. When San shamans enter trance, they are, as we have seen, said to 'die'. Richard Katz put it like this: 'Although the Kung distinguish between final death, when the soul permanently leaves the body, and the death of kia [*/aia*, 'trance'], when the soul goes out but hopefully returns, there is only one *experience* of death, and the experience is what matters' (Katz 1982:116, also 45, 99–100; original emphasis). A direct association of opening and closing eyes with trance experience is explicit in a description of a shaman that Katz (1982:99) recorded: 'His face is also blank; his eyebrows might be slightly raised, and his eyes might close and open'. The little Lion was alarmed because he realised that the Mantis was exercising his powers to enter trance.

We can now see that the 'winking' incident ties in with the other components of the central episode: getting into a bag, covering with a kaross, 'whispering' and transformation into a hare. All are in some way associated with the activities and experiences of San shamans. The non-sense elements in the tale thus relate to one another and combine to make a powerful statement.

Lions

Together, these happenings set the stage for what we began to see in the previous chapter is a recurring element in San shamanistic beliefs: conflict in the spirit realm. To approach it in this context, I note that there is evidence from all San groups for a strong association between shamans and lions. Although benevolent shamans are believed to have the power to transform themselves into lions, leonine shamans are usually considered dangerous (Katz 1982:227). The Lions, the Mantis's affines, are probably also a San category of beings: shamans (dead or alive) in feline form. Some non-carnivores are also believed to have shamanic powers.

Elizabeth Marshall Thomas (2003), who lived for extended periods in the 1950s with the Kalahari Ju/'hoan San as part of the Peabody Harvard Southwest Africa Expeditions, found that relations between San groups and prides of lions can vary. In some circumstances the relationship is one of mutual respect and general avoidance, leopards being considered a much greater danger than lions. In other circumstances, however, especially if domestic animals like cattle and dogs are present, people can be markedly afraid of predatory lions. Marshall Thomas also found that the Ju/'hoansi accorded lions some of the same attributes that they accorded the malevolent spirits of the dead:

> As during trance-dances, trancing people would confront the //gauasi [spirits of the dead], so too would they confront lions, running out into the darkness *while in trance* for the purpose of encountering lions whom they would then vilify verbally. It was my very strong impression that on these occasions lions were not actually present, or not very often, but were believed to be aware of the trancers, just the same. (Marshall Thomas 2003:74; emphasis added)

Comparable beliefs existed in the nineteenth-century south. When a /Xam shaman became very violent in trance, lion's hair was said to grow on him: 'Lion's hair comes out on his back, people rub it off with fat. . . . Then the man leaves off beating [*!khauken!khauken-ī*], when the lion's hair has come off' (Bleek 1935a:2). In other words, he changed into a lion as he was trembling in trance. Antelope fat, a potent substance, was applied as an antidote to prevent transformation into a dangerous creature. A Ju/'hoan shaman described what it was like to witness others transforming into lions. Keeney (1999:81) paraphrased his words: 'Their ears change shape and they grow lion's fur. Their hands become identical to a lion's paws and they roar mightily.' Another man explained from personal experience what this transformation feels like: 'When I start to become a lion, I feel pain and start to cry. I have to leave the dance and go into the bush where I make the change. The lion's spirit changes my mind and body. Fur grows out of my skin and claws grow from my hands. This is when I am most powerful' (Keeney 1999:93). These frightening changes, it should be noted, cannot be seen by everybody: 'When a healer [shaman] changes into a lion, only other healers can see him. To ordinary people, he is invisible' (Katz 1982:227).

Also in leonine form, benevolent shamans are believed to be able to drive off malevolent shaman-lions and thus protect people at night. A Ju/'hoan man described these antagonists as 'lions of god' and added that 'they were real lions, different from normal lions, but no less real' (Katz 1982:115). In a graphic turn of phrase, a !Kõ man claimed that a shaman in feline form is able to 'mix with' a pride of real lions without fear (Heinz 1975:29). This ability explains why the Ju/'hoansi use their word for 'pawed-creature' (*jum*) to mean 'to go on out-of-body travel in the form of a lion' (Biesele pers. comm.). So powerful was a man in feline form that the /Xam believed that if a lion was wounded out on the veld and did not die, it was really a shaman (Bleek & Lloyd 1911:187). These pervasive ideas also feature in San trance dance experiences of battles between beneficent shamans and evil, feline shamans: 'The great healers [shamans] went hunting as lions, searching for people to kill. Then someone would shoot an arrow or throw a spear into those healers who were prowling as lions' (Katz 1982:227; see also Chapter 9).

Eyes

The pervasiveness of all these beliefs about lions and shamans suggest that the Lions in the /Xam myth should probably be seen as potentially malevolent shamans *in addition to their manifest status as affines*. In this fraught context, it seems likely that the Mantis's curious threat to poke out the little Lion's eye may be seen in terms of this sort of preternatural shamanistic conflict. Eyes are clearly important beyond the blinking I have discussed. When he is put in the bag, the Mantis himself insists that he will be able to 'look out with my eye'. Later, he 'kept peeping out', and the Little Lion 'saw the Mantis's eye as the Mantis looked out from inside the bag' (Bleek 1924:16). Our next step is to seek out the San connotations of eyes and what eyes can do. A number of different contexts need to be examined.

When coming up to a dead eland, /Xam hunters avoided looking at it: 'For they are afraid that their eye might by looking make it lean' (Bleek 1932:239). In another hunting practice, a man who had killed a baboon cut fine lines round his bow to prevent the 'baboon's eyehollow' from being in his bow and, presumably, disrupting his future hunting (Bleek 1931:174). In another context altogether, worried people look at an approaching thunderstorm: 'We look, making its thunderbolts turn back from us; for our eye also shines like its thunderbolts. . . . It appears to fear our eye. . . . It respects our eye which shines upon it' (Bleek & Lloyd 1911:397). Diä!kwain explained the logic (a shared property) behind this belief: he said human eyes shone like thunderbolts. That is why the human eye can control storms. We are reminded of the Meerkats' shiny 'buttons' (Chapter 4). In an especially striking instance, Lloyd learned that if a girl at puberty (a time when she possessed the potency of the rain and was considered dangerous) looked at a man, he could be turned into a tree: the 'maiden looks at him, with the maiden's eye; the maiden looks fastening him to the ground' (L.II.2.296). Together, these instances, ranging from hunting to weather control to the powers of new maidens, show that eyes were believed to have special powers.

Similar significances were reported about /Xam shamans (who 'resemble lions') and their eyes. /Han≠kass'o said, 'A look it is with which a sorcerer [*!gixa*] takes a person who is good-looking, holds him with a glance. Then that man falls ill' (Bleek 1935a:7). Dangerous power was associated with a shaman's eye. In addition, we have already found that Ju/'hoan shamans speak of sending their eyes along 'threads of light' to watch a dance in another village (Keeney 2003:42). San shamans' eyes have preternatural powers, and they themselves are seers, in the literal sense of the word.

All in all, the poking out of the little Lion's eye meant much more to the /Xam than it does to us today. Almost certainly, it signified more than simply infliction of blindness and, perhaps, physical death. Rather, it meant destruction of the little Lion's potential shamanistic ability to 'see' into the future and to far away places and, especially, to use his eyes to inflict harm.

This seemingly small incident in the narrative is the climax of conflict. After having 'winked' and having turned himself into a hare, it was the Mantis's really serious threat of poking out the little Lion's eye that provoked the Lioness to retaliate: she stamped up to him and stood on him 'to crush him' (Bleek 1924:16). Here the Mantis seems to be conceived as an insect, a *Mantis religiosa*, that can be crushed. We are moving further and further away from the 'realistic' San person that the Mantis was at the beginning of the tale.

Flying and Dying

The insect transformation is not, however, indicative of vulnerability. Instead, it facilitates the Mantis's escape, for he then 'got feathers' and, accompanied by his possessions, 'flew up into the sky'. The 'feathers' may refer to a praying mantis's ability to transform from what is apparently a lifeless stick into a flying creature with large, spreading wings and that is itself a hunter and carnivore. In some species of mantis, this startling transformation may be one of the reasons why the /Xam trickster is called /Kaggen, the protean Mantis. Be that as it may, flight is a universal image of trance experience: worldwide shamans interpret the neurologically induced sense of weightlessness and changes in perspective that certain altered states of consciousness create as flying to distant and strange realms.[3] A Kalahari San shaman graphically reported this experience: 'When I dance and the ceremony gets very hot and serious, I feel my body fly up into the air. I float up to the height of the people' (Keeney 1999:93). A blind Kalahari shaman gave this description: 'I become so tall that I see the people as small, as if they are standing far below me. It's like I am flying over them. . . . I can even fly to another village and dance with the people there. It is only when you sing a lot and get very hot that your spirit can get up and go away' (Keeney 1999:61).

Having flown up into the sky, the Mantis descends into the water. His descent seems closely linked to his flight because the Ichneumon, in his retelling of the incident, says, 'The Mantis was in the water, *because* [he] had flown through the sky. *Therefore* he went into the water' (L.II.5.540–41; emphases added). The unity of flight and diving into water is again evident in 'The Mantis and the Cat'. The Mantis escapes from the Cat in the same way:

'The Mantis quickly got feathers, he flew away. . . . The Mantis flew along the sky; he flew into the water, he splashed in the water, he jumped out' (Bleek 1924:19). The same set of ideas, though in a different sequence, was expressed by a Ju/'hoan shaman who spoke of entering a subterranean stream and then climbing up into the sky until he reached god's dwelling (Biesele 1980:54–62).

Being underwater is a San way of speaking about trance experience. It entails weightlessness, difficulty in breathing, affected vision, a sense of being in another world and inhibited movement (Lewis-Williams 1980; Lewis-Williams & Dowson 2000:54–55). Clearly, we are dealing with a set of interrelated experiences. As was implied in the creation of the eland (Chapter 3), a waterhole was an important place for the San, a portal between realms.

The Final Episode: Restoration

When the Mantis arrives home, he causes his wife and daughter great distress by telling them that the Lions have killed /Kwammang-a and the Ichneumon (Bleek 1924:18). In Dorothea Bleek's published version this deception appears to be another example of the Mantis's antisocial behaviour. No doubt that is partly true. But, in a sentence that Dorothea Bleek altered in the published version and that recurs in the Ichneumon's unpublished retelling of the story, the Mantis explains in what is for Westerners another *non sequitur* why he wanted them to be deceived: 'I wanted /Kwammang-a to come carrying quagga meat' (L.II.5. 534, 536).

But there is another explanation. San men are required to be modest about their hunting abilities. They respond evasively when asked if they have made a successful kill, and they deliberately underestimate the amount of meat on an animal when they report a kill to their families. This custom was evident long ago in 1863. The traveller James Chapman wrote: 'When I kill anything, it is usual for our fellows, especially my Bushmen, to exclaim; "How small!" The wounded one that escapes is pronounced wonderfully fat' (1868[2]:79).

Diä!kwain explained this custom. A /Xam man returning from a successful eland hunt did not say, 'I have shot an eland'. Instead, to explain why he was limping in sympathy with the wounded antelope, he told the people 'that a bush must have pricked his foot' (Bleek 1932:234). He also avoided telling them that the eland he had shot was fat: 'He tells the people that the eland is lean, and does not tell them that the eland is fat' (Bleek 1932:239–40). The Kalahari San similarly speak slightingly of an animal they have shot; to speak positively is thought to jeopardise the quality and quantity of the meat (Lee 1969; Lewis-Williams & Biesele 1978). The Mantis's mendacity in telling the people at home that /Kwammang-a and the Ichneumon were dead and his, to us apparently contradictory, explanation that he 'wanted them to come carrying quagga meat' probably made sense to San hearers because it was a commonly observed custom.

Then, when /Kwammang-a and the Ichneumon have put down what seems to be a generous portion of quagga meat, /Kwammang-a sits in resentful silence. In accordance with San custom, he would not, as we have seen, address his father-in-law directly. The Ichneumon, on the other hand, is able to speak directly to his grandfather. He retells the story and condemns the Mantis's bad behaviour at the Lion's house and, the way he deceived his family by telling them that the Lions had killed /Kwammang-a and himself.

Who Benefits?

When the various nuggets and shamanic incidents in this tale are uncovered, a pattern is revealed. It is, I believe, this almost subliminal pattern that lifts the tale from what some scholars would deem a 'folktale' to the status of a 'myth'—if we accept the common usage of those terms. Instead of being a mix of bizarre events, perhaps a child's fairy story, the tale is a network of interrelated concepts. The cumulative affective impact of the nuggets creates a richness and allusiveness that goes well beyond entertainment (though performances of the narrative were probably that too) to the invoking of the essence of San society and its links to the encompassing supernatural realm.

Although narrators could shift emphases in their performances of the myth, the role of the Mantis in all his contrariness probably remained central in all instances. Even though his clash with the Lions comes to naught, his powers remain intact and he gets the quagga meat. There is a certain tension between the surface narrative of his misdemeanours and his supernatural powers, as they are dramatised in the nuggets. In the tale, the Mantis dramatises and reminds people of shamans' supernatural accomplishments. This means that the beneficiaries of all performances of the myth were, to greater or lesser extent, the shamans of the community: their abilities were confirmed and their reputations enhanced. Other people benefitted from the shamans' activities, but it was they who protected people against supernatural attack by spirits of the dead and by malevolent shamans who sought to shoot arrows of sickness into people. At the same time, the Ichneumon scolds the Mantis. Similarly, if people scolded a rain-maker named //Kunn, he capriciously caused rain clouds to disperse (Bleek 1933:386). He complained that people were not grateful to him. Even shamans must submit to the norms of society; people can turn on them if they fail to deliver what they promise. San social relations are complex and highly contingent.

'A Visit to the Lion's House' is thus not a simple morality tale in which good and bad are clear-cut, easily recognisable entities. It is closer to real life with all its blurred morality; it is situated in the tensions created by cross-cutting loyalties. If we insist on a moral component to San tales (why should we?), we can say that we have here a morality in which a person who could do good by contacting realms beyond the reach of ordinary people could nevertheless be taken to task by dissatisfied human beings. One who had the power to make or withhold rain, who could heal the sick but who could also turn into a marauding lion seems very human. Readers of San folklore will have their own lists of favourite tales, but for presenting the ambivalence of good intentions this is surely near the top of mine.

In the next chapter quaggas reappear, still in a role as prey, along with the Mantis and some other, on the face of it unlikely, species. There are different characters, but are they different stories?

1 The vast, semiarid plains where the nineteenth-century /Xam San lived. They made rock engravings on the black dolerite rocks in the foreground.

2 The high Maloti-Drakensberg mountains that Joseph Orpen crossed to reach the area where Qing lived. Shelters under the rocky ridges were living places for the San. It was there that they made rock paintings.

3 A chromo-lithographic reproduction of the portrait of //Kabbo that Wilhelm Bleek sent to Karl Lepsius. A Cape Town artist, William Schroeder, painted it in October 1872 in the Bleek family's home. Bleek described it to his cousin Ernst Haeckel, the famous German biologist and philosopher: 'An aquarelles portrait of my Bushman teacher, Old Jantje Tooren' (Bank 2006:185).

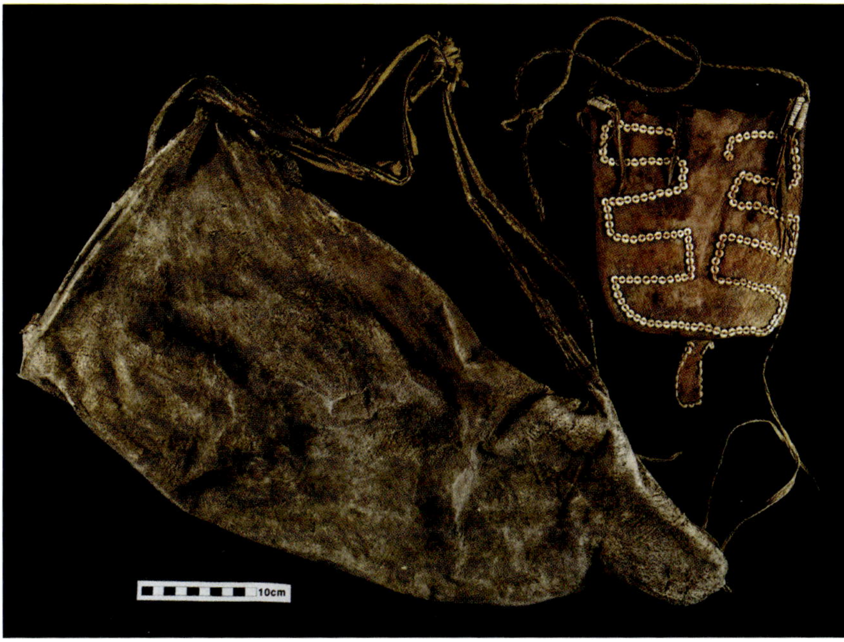

4 Kalahari San antelope skin bags. The quadrilateral general-purpose bag is decorated with ostrich eggshell beads. In some San myths, bags like this are credited with supernatural powers. The conical bag is a man's hunting bag in which he carries his bow and quiver of arrows.

5 A tracing of a rock painting showing two antelope skin bags transforming into eland. They are surrounded by realistic eland and nonrealistic creatures (copy by P. Vinnicombe, KwaZulu-Natal Museum).

6 A highly detailed running figure with many arrows. It is to the right, but part, of the section of the panel shown in Figure 7.2. The lines on its legs are not well understood.

7 San rock painting of a seated man playing a musical bow. He is part of the panel shown in Figure 7.3. The lower end of the bow rests on a resonator, probably a gourd or an ostrich eggshell; the player taps the string with a stick. A clapping figure is seated next to him.

8 San rock painting of a female eland emerging from a step in the rock face. It is realistic and delicately shaded. Note the detail in the head and the black hoofs.

9 Some of the paintings that Mapote made for Marion How in 1930. He asked for eland blood but had to make do with ox blood. The stone is preserved in the Origins Centre, University of the Witwatersrand, Johannesburg.

10 Kalahari San thumb-piano. Widespread in Bantu-speaking Africa, the thumb-piano was adopted by the Ju/'hoansi in the second half of the twentieth century.

11 A group of Ju/'hoansi people talking to Megan Biesele in the Kalahari Desert.

12 The postapartheid South African coat of arms. The central figures derive from a San rock painting. The motto is in the now-extinct /Xam San language that Wilhelm Bleek and Lucy Lloyd recorded in the 1870s: 'People who are different come together'.

CHAPTER SIX
The Mantis Dreams

I am by myself, killing to save my friends.
(/Han≠kass'o, L.VIII.17.7549)

The pattern of beliefs and rituals that we have so far traced through a series of /Xam San myths exists in numerous permutations. Each of these permutations reiterates some of the fundamental principles and conflicts that underwrite San social and religious life. Although creation of an animal does indeed play a role in some tales, as it did in the eland creation myth, we can see that much more is at stake in many of the narratives. The chief reason for the existence of the /Xam myths I have selected was, I suggest, not to provide an explanation for how the world and life in it comes to be as it is. Principally, it was to influence social relations by foregrounding the activities of shamans. This point comes out clearly in a myth that was at one time thought to be purely aetiological in that it explained how the wildebeest antelope came to have a tail like that of a horse. The comparison with a horse must, of course, date from the period of contact with European travellers and settlers, but the tale probably existed before that specific comparison was made.

The Mantis and the Wildebeest

Lucy Lloyd titled the tale 'The Wildebeest, the Mice, the Quaggas and the Mantis, or why the wildebeest has a white/light-coloured tail' (L.VIII.30.8651). /Han≠kass'o began dictating it on 31 July 1879. He said he heard it from his maternal grandmother, ≠Kammi (L.VIII.30.8650'). By 1879 he had become familiar with the tedious process of dictation and transcription, and, notwithstanding the alien circumstances, Lloyd was able to describe him as 'an excellent narrator' (Bleek & Lloyd 1911:xi).

Myth and Meaning: San-Bushman Folklore in Global Context by J. D. Lewis-Williams, 133–148

When, in 1924, Dorothea Bleek eventually edited and published the Wildebeest tale in *The Mantis and His Friends,* she titled it simply 'Why the Wildebeest Has a Light Tail'. Following is a lightly edited version made directly from Lloyd's manuscript (L.VIII.30.8651–8667; see also Bleek 1924:58–59). I have retained all repetition but have adjusted punctuation and added paragraph breaks. The title is Lloyd's.

THE WILDEBEEST, THE MICE, THE QUAGGAS & THE MANTIS,

or

Why The Wildebeest has a white/light-coloured tail

This is what the Black Wildebeest used to do. He came to one of the Long-Nosed Mice who had put ostrich feathers fastened on sticks into the ground to hunt Quaggas. The Mouse sat on the hunting-ground looking at the Quaggas. The Wildebeest unloosened the Mouse's bow string and swallowed it. He replaced it with his own entrails. I think that he had them with him so that he might give them to the Mice. The Mice were foolish people. The Wildebeest wanted the Mouse to break the bowstring when he tried to shoot. He also blunted the Mouse's arrowheads.

The Wildebeest then took a piece of a certain kind of grass and made a tail out of it. Then he went in among the Quaggas. When the Mice startled the Quaggas and the Quaggas ran, the Wildebeest ran in front of them. He came running and looking. He noticed the Mouse's screen of bushes. He ran up to it. He went and trampled on the Mouse, crushing it. And the Quaggas ran past and away, while the Mouse died.

The Wildebeest went along and took out the piece of grass. He laid it down because he was a man. Therefore he used to go and take a piece of a certain grass which becomes white when dry. He made his tail out of this grass when he intended to kill a man. When he intended to kill people, he made himself into a Quagga by means of his tail which was like theirs. This happened when he wanted to kill people.

And the Mice again put ostrich's feathers fastened on sticks into the ground so that they could hunt Quaggas. The Wildebeest came to a Mouse to blunt the tips of his arrowheads. He also unloosened the

Mouse's bowstring. He replaced it with his own entrails. He arose and took a piece of grass and made himself a tail of the piece of grass.

He then ran on in front of the Quaggas, looking as he ran. He spotted the Mouse's screen of bushes and ran up to it. When he reached the screen of bushes, he trampled upon the Mouse, trampled crushing him, while the Quaggas ran on to a place where there was no man. The Mouse lay dead. The Wildebeest took out the piece of grass and laid it down.

Then the Mantis dreamt about it. He spoke to the Striped Mouse about it and told the Striped Mouse that he should hunt at the place from which the Long-Nosed Mice did not return. The Long-Nosed Mice did not return home; they remained away. Therefore the Striped Mouse was the one who should go hunting. He would not allow the Wildebeest to remove the tips from all of his arrowheads. He should hide the arrowheads and a bowstring.

So the Striped Mouse went out hunting with the Long-Nosed Mice. I think that the Striped Mouse was the one who was clever. He went and spotted the Quaggas. The Long-Nosed Mice put ostrich feathers fastened onto sticks into the ground so that they could hunt the Quaggas.

The Wildebeest came to the Striped Mouse and told him that he should give him his arrows so that he could make them nice for him. But the Wildebeest removed the tips of the Striped Mouse's arrowheads and blunted their points so that the Striped Mouse's arrow would not enter his flesh, being deflected by his skin. He then unloosened the Striped Mouse's bowstring and strung his own entrails for the Striped Mouse. He then took a piece of grass and made himself a tail.

Meanwhile, the Striped Mouse took out a new bowstring, unloosed the Wildebeest's entrails, and strung on the new bowstring. He also took out the arrowheads, which had been inside the arrow-bag.[1] He mounted them upon the reeds.[2] He constructed a large screen of bushes which was plainly visible. He constructed a little screen of bushes which was not visible to the Wildebeest.

The Wildebeest ran on in front of the Quaggas. As he ran, he spotted the larger screen of bushes. He ran up to it, and the Striped Mouse shot him right through the armpits.[3] The Wildebeest then snatched out the piece of grass, of which he had made his tail and hurled it away. He went and lay down to die, while the Striped Mouse lay shooting at the Quaggas[4] while they ran past.

Lloyd's Notes

1. [8663'] The quivers remain at home; bags are those in which they (the Bushmen) put the arrows, put arrows which are not many into the bag. It if rains, then, they take the quiver on account of it (to keep the arrows dry).
2. [8663'] The arrowheads were fastened to the wooden shafts, ready for mounting on the reeds.
3. [8665'] The Wildebeest was a man; he made himself into a Quagga by means of his tail which was like theirs so that he might go and kill people.
4. [8666'] The Striped Mouse was a man. The Striped Mouse made his people of the Long-Nosed Mice. The Striped Mouse was strong because he was clever. The Mantis was the one who dreamt that the Wildebeest had been killing the Long-Nosed Mice, and that the Long-Nosed Mice did not return home.

Is this tale aetiological, as Lloyd's supplied title suggests? If at all, it is only marginally so. Researchers often overlook the fact that all the formal titles in the Bleek and Lloyd Archive were devised by the recorders and are frequently tendentious. /Han≠kass'o said that the grass tail was merely a temporary disguise that the Wildebeest adopted when he wanted to kill someone. The story ends with an explicit disclaimer: the Wildebeest 'snatched out the piece of grass . . . and hurled it away'. He rejected the fake tail. It is therefore only by ignoring /Han≠kasso's actual words that the story can be said to be essentially aetiological. What, then, is the myth 'about'?

To answer that question I now compare the Wildebeest tale with two other San myths, also given by /Han≠kass'o. This comparison brings out the pivotal narrative point shared by all three tales. Owing to limitations of space, I give only short summaries.

Two Parallel Myths

THE MANTIS, THE MICE, & THE BEETLE

(L.VIII.17.7542–7549; Bleek 1924:65–66)

The Long-Nosed Mice, who are people, are attacked in a riverbed and killed by the Beetle. The Mantis dreams about the Beetle and tells the

Striped Mouse to go into the riverbed to seek 'Bushman rice' (termite larvae) and to fight the Beetle. The Striped Mouse does so and makes holes for 'houses'. His wife warns him of the Beetle's approach. She calls him out of a hole that he had entered, presumably to seek 'Bushman rice'. The Beetle thinks he can trick the Striped Mouse, as he had tricked the Long-Nosed Mice, but he fails. (At this point in his performance, /Han≠kasso dramatised the Beetle's cries, 'Get out of the way / The Beetle is throwing', at some length and imitated the noise of the two different directions of the thrown spears. The Striped Mouse knocks the Beetle down. He says, 'I am by myself, killing to save my friends'. The Long-Nosed Mice, whom the Beetle had killed, come back to life. Then the Striped Mouse marries another Mouse, and the other Mice follow him.

THE MANTIS, THE LIZARD, & THE MICE

(L.VIII.30.8671–702, 31.8703–36; Bleek 1924:60–64)

Dorothea Bleek translated this tale from Lloyd's phonetic transcription in August 1919. A longer and more dramatised performance, it expands on how the Striped Mouse achieved his victory. The tale is prefaced by the Lizard's song, which is repeated in the course of the tale. The narrative itself begins when the daughter of the 'Kogelman' Lizard (a species of the genus *Agama*) alerts her father to the approach of a Long-Nosed Mouse, who is following a wounded springbok. The Lizard tricks the Mouse by asking him to enter a hole to retrieve a bag for his daughter (later said to be the Striped Mouse's daughter-in-law). The bag (and apparently other things in the hole) is said to be singing. He tells the Mouse to drink from a decayed water-bag, and the polluted water causes him to be easily deceived. He also tells the Mouse to close his eyes and turn his head away as he emerges. As he emerges, the Lizard hits and kills him. The Lizard then kills the wounded springbok and gives it to his daughter. This happens a second and a third time, and the Lizard kills two more Mice.

Then the Mantis dreams about what is happening and tells the Striped Mouse to trick the Lizard by persuading him to enter the hole. The Lizard's daughter senses that this Striped Mouse is not like the Long-Nosed Mice: he smells different. The Striped Mouse kills a springbok but declines to enter the hole and take the bag out for the Lizard's daughter. He also ignores the decayed water-bag and drinks from a new one. He then tricks the Lizard into entering the hole. While he is doing this, the Striped Mouse's daughter-in-law (the Lizard's daughter) gives the Striped Mouse the stick that her father had used to kill the Long-Nosed Mice. As the Lizard emerges, the Striped Mouse kills him. The Striped Mouse says, 'I am by myself killing to save my friends'. They carry the springbok meat back to their camp; the Striped Mouse walks with the girl. Finally, the Mantis emphasises that his dream was indeed true: 'I wanted you to see this, for you seemed to think that I was cheating you, although I spoke the truth' (Bleek 1924:64).

Initial Comparisons

There are a number of parallels between these three narratives, and indeed between some other /Xam tales (for example, Bleek 1924; Hewitt 2008:153–54). In all of them food-getting is thwarted by someone—the Wildebeest, the Beetle, or the Lizard. The Mantis then dreams and tells the Striped Mouse how to outwit the spoilers, and foraging is restored. The trajectories of the three narratives may be expressed thus:

Thwarted foraging → The Mantis dreams → Restored foraging

At once we can see an important point. Exactly what constitutes the first and last elements is not what matters most. Their content varies; the tripartite structure is merely a narrative framework. The nuggets embedded in it contain the burden of the myth by signalling its social and cognitive setting. No matter what gloss particular performances may put on the tales, a fundamental, even if muted, 'message' of the myths remains the repeated pivotal element—dreaming, and how the San understood that state (Guenther 2014; Lewis-Williams 1987, 2013a). There is, of course, no reason to suppose that all San myths follow this pattern or that they convey their meanings in exactly the same way.

The Protagonists

The following thumbnail sketches give an idea of how the San themselves regarded the principal actors in the three tales. For instance, beyond its narrow denotation, the /Xam word *!khau:/ko* meant something rather different from what 'wildebeest' means to modern English-speakers. The ethnography I now cite, coming as it does from the */Xam* and from the Kalahari San, shows that underlying beliefs about the protagonists and the intertwining of the material and supernatural realms were expressed in a variety of ways in different San linguistic communities. I omit the Beetle because little is known about him.

Black Wildebeest or Gnu (Connochaetes gnou)

Early Western travellers frequently commented on the vast herds of wildebeest that roamed the southern African plains.[1] Significantly from the point of view of the Wildebeest tale, this antelope lived in mixed herds with quagga in the past (Bryden 1893:211; Harris 1838:61). The San ate wildebeest in large numbers (Steedman 1835:(1)147; Stow 1905:85), and the missionaries Arbousset and Daumas (1846:256) recorded a prayer to the Mantis to lead a wildebeest to the hunter.

Early writers also reported that the wildebeest seemed to combine characteristics of different species.[2] For instance, Henry Bryden (1893:356, 370) noticed that it had the 'head of buffalo, body and tail of a horse, with legs of an antelope'. In the light of the San myth, it is of interest that a number of writers commented on its equine tail: 'It is long, full, and sweeping, and of a yellowish-white colour' (Bryden 1893:207).[3] The 'character' of the wildebeest is also relevant to the tale. Wildebeest are 'constantly engaged in sparring' and 'almost aggressively inquisitive' (Shortridge 1934:472, 473; cf. Bryden 1893:207). Ju/'hoansi told Megan Biesele that an angry man was said to be a wildebeest (Biesele pers. comm.; Colour Plate 11).

Like numerous other animals, wildebeest had supernatural associations for the San. In a Ju/'hoan tale, the great god ≠Goa!na becomes a little wildebeest to trick some women (Marshall 1962:232). In addition, the Ju/'hoansi place eland and giraffe fat in the zam tortoise shells that they cause to smoulder in the medicine dance, but not wildebeest fat because, they say, of its strong odour and stickiness (Marshall 1999:57). In a comparably negative context, the San of the southeastern mountains were said to place a wildebeest skull in a spring to cause it to dry up, thus discomforting their enemies (Ellenberger 1953:80). Nevertheless, the San used hairs from wildebeest tails for their paint brushes (Ellenberger 1953:87).

Quagga (Equus quagga)

The now-extinct quagga was similar in appearance to zebra, but its stripes were limited to the neck and forequarters. The /Xam used the word */habba* to mean both zebra and quagga (L.II.3.417; L.VIII.26.8282'; Bleek 1956:286). They also said that the quagga's tail was like that of a horse and hence somewhat like a wildebeest's (L.VIII.26.8282'). Its range across the subcontinent coincided with that of the wildebeest (Roberts 1951:247).

The Bleek and Lloyd manuscripts contain much about eating quagga meat, but quagga hunting could be hazardous. In an account of people driving quagga while hunting, an informant said that the animal sometimes stubbornly went in one direction. The /Xam interpreted this behaviour as leading the unwary hunter to where a lion was lying in wait (L.V.8.4578ff).

Lizard (Agama)

The male Lizard, a person of the Early Race, was said to lie in a tree and to 'bewitch' the rain, an accomplishment that makes him akin to San shamans of the rain. People shot at the lizard to make it descend. The feather of the arrow would then become a rain cloud (Bleek 1933:302). In one of /Han≠kass'o's tales the Lizard enters a whirlwind and then brings home his own flesh; his wife, deceived, thinks it is quagga meat (L.VIII.12.7114–18). Perhaps more significantly for the present context, Diä!kwain mentioned that a lizard sheds its skin, as do snakes (L.V.22.5822'). He offered this information when he was telling Lloyd that people change form when they die, as do spirit people and shamans (Bleek 1956:350; Bleek & Lloyd 1911:369). 'Shape shifting', we again see, was part of San thought.

Mice

There are many species of southern African mice; we cannot be sure to which the narrators were specifically referring, even when they refer to a Striped mouse. The Long-Nosed Mouse, however, is probably not a mouse at all but rather a species of shrew (cf. Deacon & Foster 2005:135).

It may seem odd that the selected myths present two species of mice as representative of the San. Mice are neither carnivores nor hunters. Westerners tend to see a mouse as a 'Wee . . . tim'rous beastie', not as something bold and admirable. There is, however, a possible reason for the analogy. Somewhat surprisingly, /Han≠kass'o said that a mouse resembles a lion (the archetypal hunter), because it is 'wont to wave about with its paws and it sits on its hind legs' (L.VIII.16.7527'). The /Xam also believed that the lion obtained his chest from a field mouse; therefore he has a huge roar (B.XIX.1778–1789; Guenther 1989:104–05). In a /Xam tale, a man of the Early Race who is put into a mouse skin turns into a lion (L.VIII.17.7526'–27'). The Kalahari !Xóõ San think of at least one species of mouse as a harbinger of death (Traill 2009:60) and possibly thus an effective hunter. It seems that, in San thought, people and mice were closer than may at first appear to Westerners. As we saw in Chapter 5, the /Xam saw human eyes as comparable to thunderbolts because they shared the quality of shining. The specific quality perceived by the San as being shared by two entities cannot be predicted by Westerners: it is essentially a cultural matter.

Though some mice construct shelters from twigs, they commonly live in holes in the ground.[4] //Kabbo specified: 'The long nosed mouse has a burrow little, the striped mouse has a little burrow' (L.II.3.471'). It is significant that both species of mice in the tales live underground, because /Han≠kass'o spoke explicitly of San shamans living in 'small holes, like mouse holes' (emphasis added; Bleek & Lloyd 1911:379; L.VIII.14.7276).

Like meerkats, snakes, and anteaters, all of whom feature in San myths, mice are creatures that physically move between tiered cosmological levels, as do shamans in their experiences. Entering holes in the ground is a repeated event in San shamans' accounts of supernatural travel.[5] We encountered it in Chapter 2, where the Mantis 'descended into the ground and came up again' three times, the last being in the form of 'a large bull eland'. In the context of the myths I discuss here, this feature sets mice apart from, and possibly in opposition to, wildebeest and quaggas, which live solely on the surface of the African veld.

The Mantis Dreams

As the tripartite trajectory of each of the selected myths shows, the Mantis's dreaming is the turning point: it is a central nugget. In the Wildebeest myth the word 'dream' appears only once, but /Han≠kass'o emphasised the Mantis's dream in an explanatory afterthought (note 4). In the second myth, he prefaced the story by foregrounding the Mantis's dream. In the third, he emphasised dreaming during the course of the tale and at the end. In this context, dreaming is comparable to the 'charms' in Qing's tale that the Mantis used to obtain useful information from Cogaz (Chapter 2).

Dreaming was, and for the Kalahari San still is, a mode of contact with the spirit realm. But dreams can be ambivalent.[6] For instance, the San of the central Kalahari say: 'A dream is something stupid or something that can be useful; sometimes it comes true, other times it doesn't' (Valiente-Noailles 1993:200). Despite this uncertainty, dreams can be instrumental: they can change the material world. As we have seen, //Kabbo said that he made rain by dreaming: 'Therefore I dreamt that I spoke. The rain assented to me, the rain would fall for me' (B.II.6.625). There are other instances of dreams causing things to happen. For instance, in the tale about the Mantis and the Ticks, '[the Mantis] dreams that all the Ticks' houses arise and come' (Bleek 1924:32). And they do. In the tale about the Cat, the Mantis dreams to make the day break. Probably because she did not understand how dreaming could be instrumental, Dorothea Bleek omitted this statement (L.II.5.556) when she edited the story (Bleek 1924:20). A nugget was thus lost in the process of publication. As Guenther puts it: 'Mantis's penchant for dreaming and magically manipulating his dreams link him to oneiric phenomena directly (and may explain why he has been referred to as a "Dream Bushman")' (2014:205).

Where does dreaming fit into the pattern of San thought? In providing an answer to that question, Biesele emphasises the relationship between dreaming and the trance dance:

> Dreams, trances, or day-time confrontation with the spirits are regarded as reliable channels for the transfer of new meaning from the other world into this one. . . . Though dreams may happen at any time, the central religious experiences of Ju/'hoan life are consciously and, as a matter of course, approached through the avenue of trance. . . . '(O)wners of medicine' [that is, shamans] . . . mediate to the community not only healing power but also information about how things are in the other world and how people in this world would do best to relate to them. (1993:70)

Dreaming and trance together play an important role in San life and thought. According to Ju/'hoansi, the most powerful dreams are experienced by shamans after a trance dance, the ritual in which they experience a deeply altered state of consciousness: 'The doctors may receive [power animals] in a dream or during a dance. . . . Dreams that take place after a strong dance often bring good teaching. . . . The dance makes you ready to have big dreams of the ancestors and the Big God' (Keeney 2003:58). Moreover, medicine songs, the music of which embodies the potency that dancers activate, may be given to humankind in dreams.[7] All this is not surprising, because the shifting realities and transformations of dreams parallel trance experiences. But dreams are not the exclusive preserve of shamans. They give ordinary people some idea, some small glimpse, of what shamans experience more intensely during a trance dance. In that way dreams seem to validate shamans' reports of trans-realm travel. 'Dreaming is [San] culture's glow, kept alight when people sleep at night and kindled whenever one of them, awake or awakened, tells a story—or a dream' (Guenther 2014:209).

The Bleek and Lloyd manuscripts refer to another aspect of the Mantis's dreaming abilities. As we saw in Chapter 4, in recounting the tale about the Mantis's fight with the Meerkats over ownership of a shot eland, //Kabbo says that it is by sleeping and, at the same time, trembling that the Mantis causes the Meerkats' belongings and the eland meat to be mystically transported to his own camp (Lewis-Williams 1997). Trembling (*!khauken* or *!kauken*), it will be remembered, is a San physiological reaction to trance.

Malevolent spirits are also combated through dreams and trances. Here we encounter a key element in the myths we are discussing. A Ju/'hoan shaman clearly expressed this idea: 'When you dream of spirits, they are the creatures that are trying to kill people. So you fight them and kill them. It's just like picking up this club in my hand and bashing them over the head. . . . You don't kill the spirits in the dream. You wake up first, kia [enter trance], and then you can kill them' (Katz 1982:218).

The parallels between this twentieth-century Kalahari statement and the nineteenth-century /Xam myths we are discussing are striking. The Ju/'hoan man made it clear, first, that dreaming and trance are interrelated, and, second, that shamans in trance kill enemies. For the Ju/'hoansi, the enemies are either the *//gauwasi,* the spirits of the dead, or spirits of unspecified malevolent shamans, living or dead. A central part of San shamans' *raison d'être* was their ability to protect their people by fighting and killing malevolent beings in the spirit realm (Chapter 9).

The ethnographic evidence thus strongly suggests that, in the /Xam tales although not necessarily in all environments, the Wildebeest, the Lizard, and the Beetle should be seen as marauding spirits or malevolent shamans (living or dead) bent on disrupting foraging and social relations. The tales concern just the sort of situation with which San shamans repeatedly deal in their altered states of consciousness and which they freely report to everyone after they have returned to a normal state of consciousness. The pivotal position of the Mantis's dreaming in each narrative thus points to the tales' overall conceptual context: the supernatural realm intervenes in daily life both positively and negatively, and it is the shamans who can protect people through their dreaming and trancing. Much else in the stories must be seen in terms of this overall context.

More Nuggets

In the Lizard tale a bag is retrieved from a hole in the ground. Although it is true that a San bag is (like Freud's cigar) sometimes only a bag (for example, note 1 in the Wildebeest tale), the overall context of the Mantis's dreaming suggests that, in this instance, there is more to a bag than being a mundane domestic artefact (Colour Plate 4). We saw in Chapter 5 that /Xam bags were associated with transformation and the shamanic potency that made transformation possible. The Mantis himself gets into a bag before growing feathers and flying through the sky. Bags can act as nuggets.

In the Lizard myth, the conflict over the bag is more than a fight over a mundane domestic object. Given what we know about San beliefs about bags, it was probably another fight over the possession of supernatural potency. It is intriguing that the bag in the hole is repeatedly said to be 'singing'. Given the supernatural context, this point possibly refers to the potency-filled 'medicine songs' that are central to the trance dance. But there is still more to this small incident. That the bag should come from underground is indicative of the underworld that some shamans visit before rising to the sky (Biesele 1993:70–73). As we have seen, a subterranean realm is a component of San cosmology; it is reached via holes in the ground, such as those inhabited by mice, and waterholes (Lewis-Williams & Pearce 2004a:51–53).

The use of the Wildebeest's entrails for a bowstring should also be seen in the shamanic context of the tale. Springbok leg sinew and gemsbok, eland, quagga, and hartebeest back sinew were generally used for bowstrings.[8] The making of a bowstring from intestines is, however, not as fanciful as it may at first seem. A nineteenth-century Western traveller found that the best string for a bow was said to be twisted eland gut (Borcherds 1861:111). As we have seen, the /Xam used eland bone for the 'link' in their composite arrows and eland hide for sandals; now we see that eland gut was considered desirable for bowstrings. What is at issue in the myth is the substitution of inimical Wildebeest intestines for the Mouse's bowstring. Reliance on an anatomical part of a malevolent shaman (the Wildebeest) will thwart the hunters' endeavours.

The way in which the Wildebeest is shot is not merely curious; it is highly significant: 'through the armpits'. 'Armpits' is another nugget. San shamans take the sweat from their armpits and rub it on their patients to protect them from the 'arrows of sickness' that malevolent spirits are believed to shoot at them.[9] In another tale, it is sweat from his armpit that the Mantis uses to heal his child (Bleek & Lloyd 1911:27). In Lloyd's translation, 'he anointed the child's eye with (the perspiration of) his armpits' (L.VIII.11.7011). Here Lloyd added the parenthesis to explain /Han≠kass'o's elliptical statement: for him the mere mention of armpits was sufficient to imply a major source of sweat, and sweat is the visible expression of potency.

Further connotations of 'through the armpits' are evident elsewhere in the Bleek and Lloyd Archive. In /Han≠kasso's account of digging up 'Bushman rice' (edible termite larvae) (L.VIII.10.6887', 6888), the Mantis says to the Koro-tuiten, a bird that has the remarkable shamanic ability to dive into holes that lead to 'Bushman rice': 'Thou shalt put thy hand under thy arm and smear my face'. This is done so that the Mantis will not die when he crawls into the hole to get the food. Crawling into a hole is not as fanciful as it may at first seem. Early travellers found the veld in some places pockmarked with such holes (Thompson 1827:[2]246). Locusts, another food resource, also originated in holes in the ground (McGranaghan 2012/4). These locust holes were believed to be under the control of shamans, who had the ability to remove the stones that sealed them (Bleek 1935a:10; Lewis-Williams & Pearce 2004a:55). A hole in the ground is then itself a nugget with cosmological implications.

We may conclude that, in shooting the Wildebeest 'through the armpits', the Striped Mouse was aiming at the Wildebeest's malevolent shamanic power. The nugget is incomprehensible to anyone unfamiliar with San thought and ritual. Sweat, armpits, entering a hole in the ground and shamanic activity are thus all connected in the web of San thought.

Kinship

A further element is explicit in the third tale and possibly implied in the second as well. It is one of those components the unspoken details of which San listeners would themselves have supplied because they knew the tales well. The implications were simply taken for granted. The roles played by the two species of Mice and the marriages mentioned in both tales imply the presence of kinship issues. As I have shown, numerous San myths cannot be understood without knowledge of the people's kinship system and its expression, together with joking and avoidance relationships, in the Mantis's family (Chapter 3).[10] Kinship is the basis for access to resources—a key part of the San social relations of production (Keenan 1981; Wilmsen 1989:162–63). Sometimes these relationships seem somewhat ambiguous, as at the end of the Lizard tale, but this did not trouble the San: they could live with ambiguity and vagueness; it is, after all, the essence of social life (Beidelman 1993:191).

Nevertheless, the central relationship in the Lizard tale is not vague. The Mouse's daughter-in-law betrays her father by giving the stick to the Striped Mouse. The ambivalent social position of San affines in ambilocal marriages and in uxorilocal bride service and subsequent virilocal marriages is implied, as it is by /Kwammang-a's presence in the Mantis's camp. The length of bride service that a young man has to perform can become a source of dispute, and marrying the 'wrong' people can be stressful, even disastrous, as we saw in Chapters 3 and 4, and as other /Xam San tales show.[11]

In situations of affinal tension, the trance dance is well recognised as a defusing ritual. People value shamans precisely as those who rid communities of disruptive tensions (for example, Biesele 1993:78). Guenther concluded that, among the Nharo, 'the trance curing dance can also be regarded as a more or less overt device for ludic conflict resolution' (1999:37). He went on to specify the occasions when a dance is called for. They are illuminating and fit in with the myths we have discussed: 'A dance may be performed after a big kill (the division of which could cause friction), after a contentious session discussing marriage gift exchanges, and sometimes at the end of a fight.' Richard Lee similarly found that 'the trance dance that sometimes follows a fight may serve as a peace-making mechanism where performers give ritual healing to persons on both sides of the argument' (1979:377). Megan Biesele describes just how this dissipation of tensions works:

> If there is bad feeling between two men, others will contrive to put them next to each other in the dance: participating in the form of brotherhood paves the way for brotherhood to re-establish itself. Sometimes, an unspoken tension in the community is voiced by the trancers and in this way its threat to harmonious relations is dispelled. (1993:78)

An 'Incantation'

Another implication emerges from the two supplementary tales. /Han≠kass'o uses exactly the same words in both tales to characterize the Striped Mouse's triumph. In the Beetle tale, the Mouse says: 'I am by myself, killing to save my friends' (L.VIII.17.7549; Bleek 1924:66). In the Lizard tale, he also says: 'I am by myself killing to save my friends' (Bleek 1924:64). But here Lloyd's manuscript translation of the same /Xam words (*//nă/nă si /khă kku !hu /ken ggú*) contains a gloss: 'I am myself killing by a blow, freeing my friends' (L.VIII.31.8730). The Beetle tale was recorded on September 21, 1878, the Lizard tale on August 6, 1879. The long gap between the two recordings and the fact that /Han≠kass'o nevertheless used the same /Xam words suggest that they were formulaic, that they may have constituted a /Xam saying or aphorism (cf. Hewitt 2008:40).

Indeed, Biesele (1996:154) writes of what she calls an 'unvarying incantation' among the Ju/'hoansi. It is: 'Let the dark of night fall!' These Ju/'hoan words are not a superficial observation. Elliptically, they relate to conflict in the trance dance between light/men and lions/darkness (Biesele 1993:114). Similarly, /Han≠kass'o's repeated phrase was probably a /Xam 'unvarying incantation' concerning shamanic protection. In trance, a shaman will kill malevolent spirits to save his friends.

Protecting the People

What, then, was the relationship between the Mantis, the original San shaman, and human shamans who told and heard tales like these? /Xam shamans performed many of the same tasks as the Mantis. Some, however, went further and aligned themselves more explicitly with the Mantis himself. /Han≠kass'o, for one, said: 'My father-in-law //Kabbo has mantises [*/ki /ka /kaggen*], he was a mantis's man [*/kaggen-ka !kwi*]'. */Ki* means to possess or own, but the word is also used to denote the special relationship between a shaman and the creature he or she was believed to control; there were, for example, 'shamans of the springbok', who could cause them to run into the hunters' ambush. The second part of the statement, where 'mantis' is given in the singular, suggests that //Kabbo had some sort of relationship with the Mantis, not just praying mantises. This personal relationship is strongly implied by what follows in the manuscript. It is an account of the Mantis's special affection for the eland and the ways that he contrives to effect the eland's escape from hunters. This passage begins: 'The Mantis does not love us, if we kill an eland' (Bleek 1924:12).

In the three selected myths I have discussed in this chapter the Mantis, an unpredictable trickster, is also a *deus ex machina,* resolving social tensions and combating malevolent spirits—something San shamans routinely do in both the trance dance and their dreams. Performances of the myths would have reminded people of the malevolent spirits that constantly try to disrupt their lives and frustrate foraging, but—at the same time—they foregrounded the shamans' power to combat those spirits. Shamans were thus architects of a controlled relationship between danger and safety, strife and amity, life and death, and, ultimately, the supernatural and material realms. Far from constituting a religious add-on to life, they were a cornerstone of San conceptual, social, and economic life.

CHAPTER SEVEN

Narrating and Painting

Kaggen makes the pictures.
(D. Bleek: A3 004:188)

Apart from the thousands of rock art images of eland—in almost all parts of southern Africa it is the most commonly depicted species—the protagonists in the myths I have discussed were seldom represented (Deacon 1994, 2001; Guenther 1994). There are, for example, very few images of wildebeest across the whole subcontinent despite their former abundance in vast ubiquitous herds.[1] In an attempt to explain this apparent anomaly, Patricia Vinnicombe (1976:210) argued that the wildebeest symbolised 'hindrance and interference, or that which thwarted plans and sapped strength'. As the preceding chapter shows, this interpretation of the possible symbolism of wildebeest is not without merit, but it leaves much of the myth unexplained, such as the role of the Mantis, dreaming, the bag in the hole, and the manner in which the Wildebeest was shot 'right through the armpits'. Nor does it explain why other species, such as mice, mantises, and meerkats, are never, as far as I know, convincingly depicted. The straightforward idea that the rock art 'illustrates' the myths is clearly not viable. Nor is it a 'record' of daily San life, its humdrum affairs and occasional excitements; that idea leaves too much unexplained (Lewis-Williams 1981, 2014).

Are the images then purely *objets d'art* with no deeper meanings (in itself an unlikely concept), as some researchers have argued?[2] To be sure, the images are often strikingly beautiful, but that does not explain why particular subjects were depicted and others not, or the many curious characteristics to which I refer in this chapter. One could argue that the beauty of the art has diverted modern people's attention from its complex role in San communities. Then

Myth and Meaning: San-Bushman Folklore in Global Context by J. D. Lewis-Williams, 149–172

again, the seemingly prosaic, even playful, nature of numerous San narratives seems to many present-day viewers to be a characteristic of their art as well. As a result, well-delineated animals and fleet human figures have been appropriated to make amusing or elegant motifs for table mats, fabric designs, and so forth. These trivial but unfortunately ubiquitous modern contexts diminish many viewers' understanding of the art.[3] On the other hand, some viewers note the seeming diversity of the images and assume that San rock art was a simulacrum of multipurpose Western image-making, some images being made simply for pleasure or amusement, others for instruction, others as mythical mnemonics, others as *art pour l'art*, and so forth. This view, always taken as axiomatic and never substantiated, robs the art of its intrinsic San-ness.

Still, there remains among researchers, and indeed all who admire San rock paintings, a persistent feeling that the images must have *something* to do with myth. They are surely right—at least in a broad sense, as I show in this chapter. To discern the link between San myth and art we must see past the narrative element in both. The link between the two modes of expression is more subtle than most Western viewers realize on their first encounter with the art. It had its own *raison d'être*.

Notwithstanding the apparent realism and beauty of many images, ethnographic and painted evidence assembled over the last thirty and more years shows that the practice of making them was essentially concerned with different types of contact with the supernatural realm and its beings.[4] San image-making was a ritual practice in its own right, not a secondary, merely illustrative appendage to San mythology or anything else.

As with the myths I have discussed in previous chapters, we need to set aside Western expectations and values and examine elements of the art in the light of specifically San beliefs and practices. Only then can we begin to discern parallels between San mythology and image-making and begin to form some idea of what the art meant to the San (Lewis-Williams 1995).

Painted Nuggets

Scattered among the images are easily missed painted nuggets, just as there are lexical nuggets in San myths. This principle is the chief parallel between San myth and imagery: powerfully allusive images and features of images situate the art conceptually. Even as verbal nuggets transform apparently mundane narratives into highly significant myths, painted nuggets transform what could be mistaken for prosaic depictions of daily life into richly evocative manifestations—not just statements—of beliefs, experiences, and social relations.

In preceding chapters, I have mentioned some of these painted nuggets in passing; they and others have been discussed in detail and illustrated in numerous publications.[5] Here, my task is two-fold: to relate mythological nuggets to their painted counterparts and to identify additional nuggets that occur only in the art. Like mythological nuggets, each painted nugget to which I draw attention is authenticated by San ethnography. We do not have simply to guess at their meanings.

Before exploring specific panels, or clusters, of images to see how they 'work', I list some painted nuggets with brief explanations. Cumulatively, they begin to give a general flavour of how the often complex and seemingly jumbled panels of images functioned as open-ended networks of meaning rather than finite, bounded 'compositions'. This fragmentary mode of 'composition' is an initial, overarching parallel between San myths and rock art.

- Dancers in distinctive postures that can still be observed in Kalahari San trance dances occur frequently.[6] They are often depicted bending forward at an acute angle and supporting their weight on one or two dancing sticks, as trancers do when their diaphragm muscles painfully contract.[7] Others hold their arms in a distinctive backward position that Ju/'hoan shamans have said some dancers adopt when they ask god for more potency.[8] In the painted panels, dancers appear in circular or scattered groups, in lines that are sometimes called 'processions' or singly and seemingly unrelated to adjacent images.
- Sometimes dancing figures are accompanied by seated, clapping women with their fingers splayed, as at real-life San trance dances.[9] Like the dancers themselves, these clapping figures appear in multicomponent dance scenes but also occasionally scattered separately in complex panels. Trance dance songs, sung and rhythmically accompanied by clapping women, contain the potency that the men harness to enter the spirit realm. Clapping figures therefore signify the activation of potency. In addition, some San women became shamans.
- Blood is frequently depicted falling from the noses of dancers and sometimes from the noses of isolated standing, walking, or running figures. As we have seen, nasal bleeding is a San physiological reaction to trance;[10] it places the relevant figures together with their sometimes less

explicitly painted companions in the domain of supernatural contact and experience. Blood also sometimes falls from the noses of supernatural and natural animals and part-human, part-animal creatures.

- The multitudinous depictions of eland point not only (if at all) to abundant food but also to all the ramifications of the eland creation myths (Chapter 3).[11] Tighter focus is often provided by painted contexts: people, some transformed, shown dancing next to eland recall the San practice of absorbing the potency that is released by dying eland. Many images show eland in dying postures.[12]
- Therianthropic figures are a well-known feature of San rock art.[13] Their largely human bodies often have an antelope head and, somewhat less frequently, hoofs. These images have been thought to depict spirits of the dead,[14] but other features of them suggest rather that they depict shamans (living or dead) partially transformed into animals. Many bleed from the nose. In addition, many have their arms in the backward or extended position that is characteristic of the trance dance. Some wear long karosses, which, as we saw in Chapter 5, were in some circumstances associated with shamans.
- Depictions of bags sometimes appear singly among seemingly unrelated images; they are not whimsically selected items of daily use. As I pointed out in Chapter 5, certain myths show that bags were associated with transformation and supernatural potency.[15]
- Flywhisks are used by the San only in trance dances.[16] In the early 1960s Richard Lee found that, among the Ju/'hoansi dancers, flywhisks were 'indispensible' (Lee 1967:31). A Kalahari shaman usually has only one flywhisk, but some painted human figures are shown with unrealistically numerous flywhisks. They are also depicted in isolation and, occasionally, in isolated bunches.
- Rain-animals, the imaginary quadrupeds that southern San shamans of the rain captured, led across the countryside, and then killed (or milked) to make rain are also depicted.[17] Often they are being led or driven by shamans. Sometimes they are shown wounded or being killed. Snakes with antelope heads are also probably 'rain-creatures'.
- Meandering red lines, often fringed with small white dots, sometimes weave thorough complex panels of many images, especially in the southeast mountains. They enter and leave images and also seem to penetrate the rock face. Variations of the form are known. They are almost certainly the 'threads of light' that San shamans routinely report seeing

in their trance states (Chapter 2).[18] Sometimes the lines simply end on a flat surface and then reappear a short distance away. Sometimes shamans are shown walking or dancing along the lines. They probably facilitated shamans' access to the spirit realm.

- Flecks of paint depicting the potency and sickness that only shamans can see are sometimes scattered around dancers and other images.[19] Some of these flecks are probably akin to, or actually represent, the 'arrows of sickness' that malevolent spirits shoot at people.
- One of the most significant nuggets is that some images (often but not exclusively 'threads of light') are painted to give the impression that they are entering and leaving the rock face via cracks, steps, or other inequalities. For the San, the rock face was not a meaningless *tabula rasa* on which 'artists' could paint whatever took their fancy but rather a deeply meaningful and contextualising 'veil' between material and supernatural realms.[20] Consequently, whatever they painted on the rock face was set in a spiritual context.

Many of these nuggets were created by synecdoche: dancing postures, nasal bleeding, clapping, flywhisks, and so forth may be features of isolated human figures and not be part of a complete dance scene. Part of a whole signals the whole, or an aspect of it (Lewis-Williams & Pearce 2004a:98–100), just as a word in a myth (for example, 'tremble') may point to a ramifying complex of beliefs. This type of compression is characteristic of San myth *and* imagery. Then, too, a nugget (say, nasal bleeding) may be associated with one human figure that is part of a line of eight or more apparently walking figures. The single figure situates the whole group as having some connection with trance activity. For images to exercise their full effect they, like myths, assume much knowledge on the part of viewers.

Overall, and whatever additional meanings the images may have had (Lewis-Williams 1998, 2001a), more paintings in the southeastern mountains point to shamanistic beliefs and practices than to any other San area of belief.[21] If other rituals and beliefs led to the making of images, those images are few, certainly in the southeastern mountains.

To exemplify the way in which nuggets appear in the art I consider three contrasting panels of images: the first is a scatter of what appear to be unrelated images, mostly realistic; the second is a clearly nonrealistic combination of nuggets; the third seems to be a realistic narrative scene but has destabilising nonrealistic elements.

Images of Quadrilateral Bags

Bags, a feature of my first panel, are one of the clearest nuggets that are common to myth and rock art and so make a useful starting point. Often people are depicted carrying bags, either the quadrilateral, general-purpose bags used, for example, by women gathering veld foods or the conical kind in which men carry their hunting equipment (Colour Plate 4). Often a painted bag is probably just a bag, a piece of daily equipment, but images of them are also scattered individually among what seem to be unrelated paintings, or they appear in curious, supernatural contexts.

In the panel shown in Colour Plate 5, two leather bags of the quadrilateral kind are in the process of transforming into eland.[22] Blood falls from the nose of one of these eland-bags, and a bow protrudes from its 'shoulder'.

As is often the case with this sort of painting, the eland-bags are set in the midst of a panel of diverse images of antelope. In this instance, there is one realistic, semirecumbent rhebuck, and as many as ten beautifully drawn depictions of eland. Three of them have their heads lowered in a posture associated with dying (Lewis-Williams & Dowson 2000:50–53). Previous chapters have touched on the supreme importance of eland for the San. As I show later in this chapter, the many apparently realistic depictions of eland, previously thought to depict objects of the hunt or to be purely aesthetic creations, were (at least in part) 'reservoirs' of the potency that San shamans activated. People, perhaps not only shamans, drew potency from those images, even as they did from real eland, when, after a successful eland hunt they danced next to the carcass.[23]

In contrast to the apparent realism of these antelope paintings, there are others that seem to fall into a different conceptual category. For instance, in the lower right of the panel the forequarters of an eland is painted to give the impression that it is emerging from a step in the rock face; the main body of the eland appears to be 'behind' the rock and so out of sight. There are also four flying creatures with their arms in the backward posture associated with trance experience. One has blood falling from its nose and is thus clearly associated with trance experience. Widely depicted, this sort of image is known as 'flying buck', *ales*, or 'trance buck'. There are also three strange quadrupedal creatures with stylised human heads; two of them have red lines across their faces, probably an indication of nasal blood. One of them also has red lines across its legs. They are clearly fantasy animals, perhaps rain-animals, although exactly what they signified is presently unknown.

The more one examines what may, at first glance, appear to be simply a herd of eland, the more one realises that the panel is a congeries of nuggets, the meanings of which play into one another. Dying eland releasing their potency, shamans using that potency to transform, transforming bags, strange creatures encountered in the spirit realm, traffic through the rock face are all features that point to the realm of shamanic experience.

How was this diversity within overall unity constructed? Although we cannot be absolutely certain, it seems that the panel was built up by a number of painters working over an unknown period of time. Unfortunately, it is not possible to determine the sequence in which images were added; their degree of preservation is no sure guide, and there is no significant superpositioning in the panel. What could have been the significance of this fundamental, cumulative feature of San rock art? Each added image seems to have contributed to and enlarged on the overall spiritual significance of the panel as a whole. I suggest that adding images to a panel was like the telling of a myth, each performance having its own nuances, or the reporting of trance experiences after a dance.

Indeed, Megan Biesele (1993:76–77) has explicitly linked the narrating of myths to the recounting of trance experiences: both, she says, deal with 'deep, long-term underlying assumptions' and are characterised by 'similar values and images'. She adds: 'A high degree of stereotyping is present in the verbal accounts of travels beyond the self which are made after a night's trancing. Yet Ju/'hoansi themselves treat these experiences as unique messages from the beyond, accessible in no other way save through trance' (1993:72).

Biesele's observations give us a major insight into San rock art, its mixture of stereotypical and idiosyncratic images. Although, to us, apparently repetitive, each image was probably a 'unique message from the beyond', some even being depicted as emerging from behind the rock face. Seen in this light, panels crowded with accumulated eland images must have been immensely powerful.

Conical Bags

In another painted panel, two bags of the men's conical kind are shown transforming into the sort of bizarre, composite creatures that are associated with trance visions (Figure 7.1).[24] Both creatures have blood falling from their noses and streaming back across their faces. They also have 'tusks' a feature that is not fully understood but that appears on numerous transformed images.

In Chapter 5 we saw that getting into a bag was like 'getting into an animal' and inhaling its potency. In Figure 7.1 the two transformed bags are juxtaposed with apparently realistic eland; one is recumbent. The most remarkable feature of the panel is that depictions of eland in a typical hunched sleeping posture have been painted, each in two parts, to give the impression of being *under* the transformed bags. The halves do not quite match up. The sleeping elands' horns and ears are just visible at the bizarre creatures' humps. The sleeping eland probably point to shamanic dreaming of the kind we encountered in Chapter 6 and are therefore consonant with the transformed

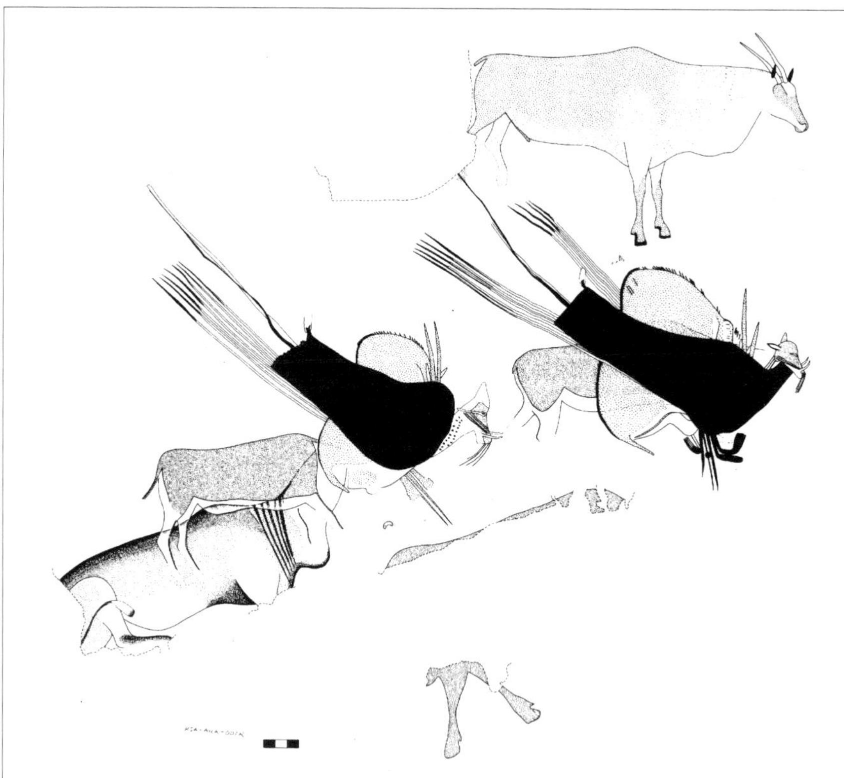

FIGURE 7.1 A tracing of a San rock painting showing two conical men's hunting bags transforming into bizarre creatures. Both bleed from the nose. Two curled-up, sleeping eland are painted in halves to give the impression that they are under the bags. (scale in centimetres)

bags. As we saw in that chapter, transformations occur in dreams, as well as in trance dance experiences. This is a case of what I call 'factitious superpositioning': images are occasionally painted in such a way as to suggest that they were made before and are therefore under other images. This curious technique shows that superpositioning, a common feature of San rock art, was meaningful: quantitative studies have shown that it was a form of syntax, not merely a result of ignoring the work of previous painters (Lewis-Williams 1974).

The composite creatures/bags themselves have numerous points of interest. Sectioned arrows of the kind we encountered in Chapter 4 and unstrung bows protrude from the bags and from behind them. San men carry their bows unstrung in conical bags to prevent the staves from weakening. Part of the left-hand figure is poorly preserved, but flywhisks (used only in the trance dance) clearly protrude from the right-hand figure's chest, along with a digging stick weighted with a spherical bored stone, a women's artefact (Bleek & Lloyd 1911:361) that was also used to contact the spirit realm. Women wishing to summon the spirits beat on the ground with bored stones (Bleek 1933:390; 1935a:41; 1936:135). What may at first glance appear to be an unlikely mix of artefacts (men's flywhisks and a woman's digging stick) in fact constitutes a meaningful group that relates to contact with the spirit realm as a result of cooperation between the sexes—as does the trance dance itself (Biesele 1993:97–98).

Biesele's perceptive insight into the relationship between contributions to the folklore tradition and 'hallucinations of actual *n/omkxaosi*' is key to understanding San rock art (Biesele 1993:76). I suggest that the repeated painting of very similar eland in conjunction with idiosyncratic images should be seen in a comparable way. To us the 'ordinary' images may be stereotypical and redundant, but the San probably valued each as an important authentic statement. Folklore, trance experiences and rock art all 'embody a kind of timeless, anonymous truth at the same time as they are being creatively renewed by individuals' (Biesele 1993:76–77).

Potency and Conflict

Unlike the transformed bags and sleeping eland, other groups of images seem to be realistic 'scenes'. Some of these clearly depict conflict. These so-called fight scenes have long attracted researchers' attention and are frequently interpreted as records of real-life events painted to commemorate an incident or to show those who were not present what it looked like.[25] Some fight scenes seem to depict conflict between San bowmen and other cultural groups, such as shield-bearing Bantu-speakers (Campbell 1986, 1987; Challis 2008, 2012). Although no reliable statistics are available, more seem to pit San against San.

In numerous fight scenes nuggets point to a supernatural provenance rather than to an exclusively historical record, although such conflictual experiences may, of course, arise in daily life.[26] Generally speaking, the more detail in a painting, the more information it transmits. 'Simple' fight scenes comprising only a few apparently opposing figures should therefore be seen in conjunction with the more elaborately painted ones and, moreover, in the context of the surface on which they are painted—the 'veil' between the material and supernatural realms.

In one of the best-known examples two opposing groups of human figures are separated by some unpainted rock. In the copy shown in Figure 7.2 six running figures to the left have been omitted to mitigate the effect of reduction to fit the page. Eight of the figures shown have quivers full of arrows; in some instances the quiver itself has faded over time. Two of the figures carry exceptionally large numbers of arrows (Colour Plate 6). In the right-hand group, arrows fly around. Most of the arrows have triangular iron points; others have bone points without barbs. In addition, two of the figures in the right-hand group and three on the left carry iron-bladed 'battle-axes' of a design associated with Bantu-speaking agriculturalists. In more recent decades, San living in the southeastern mountains obtained various artefacts from their

FIGURE 7.2 An extensive San rock painting of a conflict between two groups of human figures. Arrows fly among the right-hand group. Note the small flecks painted along the back of a seated man. See also Colour Plate 6.

iron-smelting neighbours (Chapter 2). On the right a figure removes an arrow head from his arm, the detachable shaft having fallen away. Farther to the right a supine figure bleeds from the head (possibly the nose) and back.

The spiritual provenance of the conflict is implied by nuggets that link up with those we have considered in previous chapters. In the right-hand group there are two seated bowmen. The larger one is in the front line of five figures that face the seven approaching attackers. He has small, nonrealistic flecks positioned along his spine. A likely explanation for this feature is recorded in both the Kalahari and the southern ethnography. In trance, a Kalahari Ju/'hoan shaman feels his potency 'boiling' up his spine:

> As the healer's num [*n/om*] is heated up to boiling, it vaporizes and, rising up the spinal column, induces kia [trance]. . . . In your backbone you feel a pointed something and it works its way up. The base of your spine is tingling, tingling, tingling, tingling. Then num makes your thoughts nothing in your head. . . . In kia, healers express the wishes of the living by entering directly into a struggle with the spirits and the lesser god. . . . Then your front spine and your back spine are pricked by these thorns. Your gebesi [diaphragm] tightens into a balled fist. (Katz 1982:95, 42, 43, 46)

The same sensation was experienced in the nineteenth-century south. Diä!kwain explained that when a novice was learning to trance he was given the nasal blood of his mentor to smell; this caused his 'gorge' to 'jump up; we shiver [*!kouky*]' (Bleek 1935a:12–13). On another occasion Diä!kwain was trying, probably with gestures, to convey this sensation to Lloyd. He said that it seemed as if the shaman's 'vertebral artery would break' and that when a shaman was experiencing out-of-body travel his 'vertebral artery has risen up' (Bleek 1935a:22, 23). Lloyd's translation of the word *!khăŭä* in these passages as 'vertebral artery' probably arises from her lack of familiarity with trance. Used as a verb (as many /Xam nouns can be) the word means 'to boil' (Bleek 1956:425) and was probably the southern San's metaphor to describe the rising sensation experienced in the spine as a shaman goes into trance, as it still is among the Kalahari San.[27]

In other paintings, the same boiling sensation is suggested by a line of small white dots placed along a figure's spine.[28] The rising of boiling potency up a shaman's spine is one of those things that can be seen only by other shamans (Katz 1982:106), a point that suggests that paintings of this kind were made by shamans who painted what they themselves 'saw' and experienced.

The painted 'flecks' along the bowman's spine thus constitute a nugget that situates the whole group of images: the fight is taking place in the spirit realm or in a liminal space—the real world infiltrated by the spiritual realm.

The arrows flying throughout the right-hand part of the scene recall a twentieth-century Nharo shaman's remarks in his description of dangerous San trance experiences: 'If *n/um* doesn't burn you, and *!kia* or the lions that your spirit merges or mingles with don't half kill you, the spirits and their arrows to whom you expose yourself at every trance will get you' (Guenther 1999:190). Comparably, in the nineteenth-century south, /Han≠kass'o told Lloyd that 'sorcerers are wont to shoot a person; we do not see their arrows coming. . . . They shoot to kill, they strike the man dead' (Bleek 1935a:5, 7). Katz adds an important gloss on these 'arrows of sickness' (1982:168). He found that the Ju/'hoansi believe the invisible arrows to have, almost to *be*, *n/om*. They are 'felt as painful thorns or needles'. These remarks lead me to believe that the painted arrows, those flying through the air as well as those in the quivers, may be seen as an equivalent of potency.

Potency is implied in a different way by a figure in the left-hand group; it has a raised knee and a pointing finger. Both these repeated postures relate to San shamanism, and both appear in their rock art. The raised knee posture is not well understood, but it may be related to the tightening of the stomach muscles in trance of which San shamans speak. The pointing finger is better understood. By pointing a finger and snapping his fingers a San shaman can send a powerful 'shot' of potency at a person. Lorna Marshall (1969:351–52) found that a Ju/'hoansi shaman should be careful not to 'point his finger fixedly at anyone or snap his fingers at anyone, especially a child. . . . "A fight" might go along his arm, leap into the child, and kill it' (*/a*; cf. Chapters 3 and 4). When teaching a novice, an experienced shaman may snap his or her fingers at the student's stomach; each snap signifies an arrow of potency that will take him closer to being a competent shaman. Once inside a person, the arrows multiply or 'give birth' (Katz 1982:46, 168). During a particularly intense episode of dancing a Ju/'hoan shaman stopped, turned, and pointed his finger at another dancer across the fire. The other dancer immediately fell over. Later the shaman repeated the action with the same effect. A comparable action was described in the 1980s by an old woman of partial San descent who was living in the present-day Eastern Cape Province (Jolly 1986; Lewis-Williams 1986). She said that San shamans simply pointed at dassies, which 'froze'. Then the men walked over to them and picked them up. The woman spoke only of dassies, but Lloyd was told that people pointed a burning stick of a medicine plant at springbok so that they would run slowly and easily fall to the hunters' arrows (Bleek 1936:146–47). In the light of these beliefs it seems certain that the pointing figure in the fight scene is sending a 'shot' of potency.

In the left-hand group three women appear to restrain two of the men. There are no women in the right-hand group. This difference suggests that the left-hand group is probably the 'home side', but there is more to it. During a dance, when the trancers become frenzied, women and other men hold them to prevent them from injuring themselves by running off into the darkness beyond the camp fire or even hurling themselves into the fire (Katz 1982:48). Marshall describes men in trance rushing around the dance circle and out into the darkness to confront the threatening spirits. When one of these men scoops up coals and throws them over himself, 'women are quick to reach out their arms to hold him' (Marshall 1969:377; also 376). In the rock painting, unrestrained men run in the direction of the fight. The women restraining the man probably represent another aspect of supernatural conflict.

Some of the more elaborately painted figures have complex lines painted on their legs. These lines are not well understood, but they seem not to be realistic. Some of the figures have broader bands at their ankles that may depict dancing rattles; in other paintings they are sometimes depicted with individual segments. Others may depict 'rings' of antelope skin or an edible root called *//gwi* (Lewis-Williams & Challis 2011:67).

In sum, we must remember that all conflict images were painted on the 'veil' between realms. Placed on the interface, these images manifest shamans protecting their own people from marauding spirits. In this way, they refer to a frequent theme in San accounts of spiritual experiences and, as we have seen, in some myths. Like depictions of bags and some other nuggets, fight scenes constitute a parallel between San myth and rock art. The painted nuggets that I have identified are interrelated, as are those in the myths that I discussed in previous chapters. They, too, add up to a complex statement.

Mythical Creation and Image Depiction

We can now go a stage further and ask: Are there any parallels between the mythical creation of the first eland and the 'creation' of eland images in rock shelters? To answer this question I consider in turn:

- the different locations of the mythical creation of the first eland and the making of rock art images,

- the constituents of creation and depiction, and
- the identities of the narrators and painters.

Locations

The apparently very different places where the myths situate the creation of the first eland (waterholes) and where the San painted eland (rock shelters) have common conceptual associations.

MYTHS In the /Xam narratives, birth imagery and the first eland's appearance out of a waterhole (Chapter 3) may suggest to Westerners familiar with sexual and psychological theories of myth that the waterhole represents a womb. Such birth imagery may well be present in San eland creation myths, but broad interpretations of this kind can divert attention from significant ethnographic details. They tell us little about how the myths were understood and manipulated by the San.

Similarly, eland behaviour may have had something to do with the notion of creation in a waterhole. When they are able to, eland frequent rivers and wetlands, where they browse on the tops of reeds.[29] The nineteenth-century agriculturalist Basotho people, who were in close contact with the San, had a song in praise of the eland. A line of it reads: 'It is a cow that conceals its calf in the unknown fords of the rivers' (Arbousset & Daumas 1846:47)—a striking parallel to the /Xam myth of the Mantis creating the first eland in a waterhole.

Over and above these considerations, the waterhole itself is, as I showed in Chapter 3, a nugget. /Xam beliefs about water, waterholes, and eland show (1) that, for them, the place where the Mantis made the first eland was a special point of breakthrough between the level of material life and a nether spiritual realm and (2) that eland, being so closely associated with rain/water, *embodied* the idea of transition.

In contrast to the /Xam semiarid plains and isolated waterholes, Qing's Maloti San eland creation myth (Orpen 1874:3–4) comes from a mountainous region with perennial streams and rivers. His account of where the Mantis makes the first eland begins with his wife, Coti, giving birth to 'a little eland's calf in the fields'. Because the San were not traditionally agriculturalists, it seems likely that 'fields' may not mean demarcated cultivated areas but rather open country, the southern African 'veld'. Having ascertained that the name of the young creature was 'eland' ('Tsha' in Orpen's orthography), the Mantis puts it in a gourd and takes 'it to a secluded kloof enclosed by hills and precipices, and [leaves] it to grow there' (Orpen 1874:3–4). In this well-watered region, the rivers that flow through the kloofs (steep-sided valleys) are conceptually equivalent to waterholes in the semiarid /Xam territory (Lewis-Williams 1981:32). Here we have a 'secluded' kloof; in the /Xam tale, the eland was 'hidden' among the reeds—hence the Mantis's call to him.

Access to a subaquatic, subterranean spirit realm via these rivers is seen in some of Qing's tales. In one, Qüuisi and, later, Cagn himself (Chapter 2) are both perilously held in a river by a 'thing' that catches hold of them from below (Orpen 1874:9). In another, Qwanciqutshaa, in some ways the Mantis's alter ego (Lewis-Williams 2013b), throws himself into a river: 'And there were villages down there'. Later, he moves along the vertical axis of the San cosmos and goes 'up to the sky' to fetch some of his belongings that he had thrown there (Orpen 1874:7). Qing also spoke of 'men with rhebok's heads' and 'tailed men, Qweqweté' (the name is obscure), who 'live mostly underwater' (Orpen 1874:10)—that is, in rivers and riverine pools.

In both /Xam and Maloti myths, then, the eland comes from, or grows to maturity in, a water source that is a place of breakthrough between cosmological levels.

IMAGES Waterholes and rivers are not explicitly depicted in San rock art. The rock shelters themselves, however, hold clues. As we have seen, some images are depicted as if they are emerging from or entering into cracks, steps, and other inequalities in the rock surface (Lewis-Williams & Dowson 1990). We need to look at some of them more carefully.

In one particularly rich instance, a man is depicted playing a musical bow while a spotted rain-animal emerges, not from a depiction of a waterhole but from an actual step in the rock—that is, *from the landform (rock shelter) itself* (Figure 7.3; Colour Plate 7).[30] This painting recalls /Han≠kass'o's description of a rain-maker who 'used to strike the bow-string, and then the clouds came up while we were asleep. . . . it rained there, poured down until the sun set' (Bleek 1933:390–91). Surrounding the bow player, people are in distinctive clapping and dancing postures associated with the trance dance; one figure has antelope hoofs and bleeds from the nose, both indications that he is in contact with the spirit realm (Lewis-Williams 1981:75–101; Lewis-Williams & Pearce 2004a).

FIGURE 7.3. A San rock painting of a man playing a musical bow as a rain-animal emerges from a step in the rock. A tall central figure bleeds from the nose and has hoofs. See also Colour Plate 7.

In another painting, rain shamans lead a rain-animal by means of a thong attached to its nose (Figure 7.4).[31] Diä!kwain described this ritual: 'Then they go and sling a thong over the water bull's horns, they lead it out' (Bleek 1933:376). In this painting, they are moving in the direction of painted arcs that represent the walls of a rock shelter with bags hanging from them. The San hung their bags on wooden pegs driven into the walls of rock shelters and sometimes painted bags clearly hanging from pegs (H. Deacon 1976:23, 47; J. Deacon 1999:52; Parkington 2008:20). The painted thong joins to the arcs. Two of the shamans leading the rain-animal bleed from the nose. One bends forward in a typical trance dance posture occasioned by the painful contraction of the stomach muscles.[32] The rain-animal and the shamans are behind the rock wall, that is, in the spirit realm. In the shelter itself, there are images of seated people, bags, cattle and, farther away, horses.[33]

In both these instances, painted rain-animals are being coaxed or led from behind the rock face in the same way and in the same ritual circumstances (altered states of consciousness) that purely mental rain-animals were led out of waterholes. Both panels suggest that rock shelters were, at least sometimes, specialised locations for rainmaking rituals (Challis, Hollmann, & McGranaghan 2013; Lewis-Williams & Pearce 2004b).

As we have seen, some depictions of apparently realistic eland were similarly painted to suggest that they are emerging from the rock face: part of the

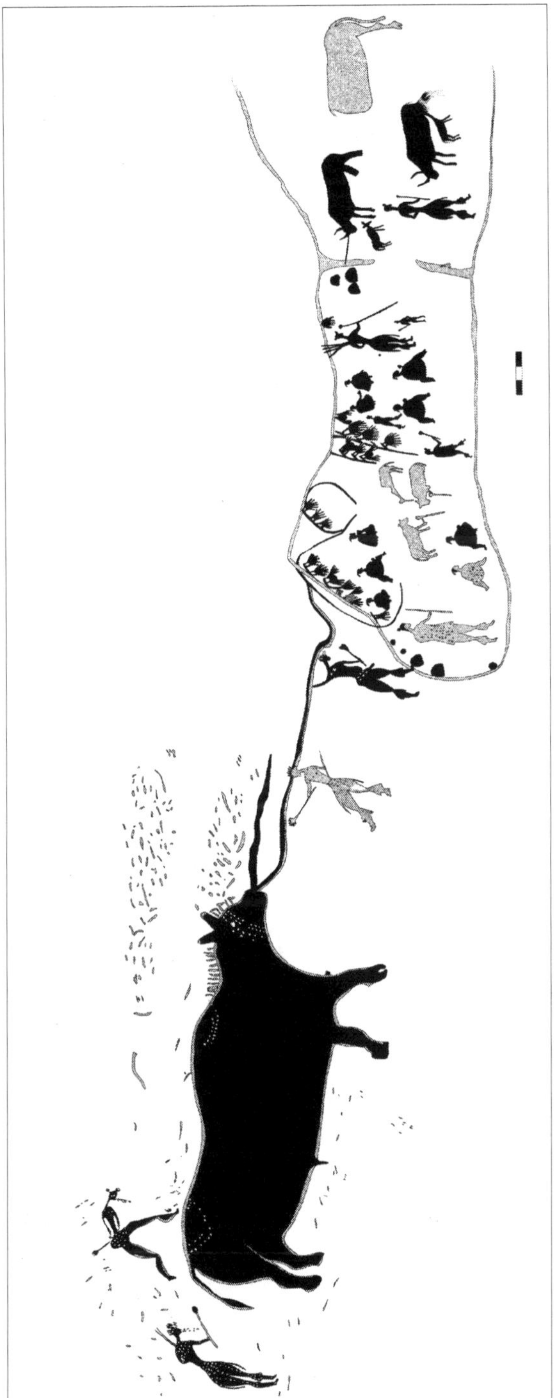

FIGURE 7.4 A San rock painting of rain shamans leading a rain-animal toward a rock shelter. Two of the figures bleed from the nose. In the rock shelter, people are seated and there are tethered cattle; farther to the right of this copy there are depictions of horses. Bags hang from the walls of the rock shelter.

antelope appears to be hidden behind the rock (Colour Plate 8).[34] It is as if the painters were 'calling' the eland through the rock in a manner comparable with the way in which the Mantis called the eland out of the waterhole (Bleek 1924:2). Other images that commonly appear to emerge from the rock face include serpents, often with antelope heads, that the /Xam associated with water and rain (Lewis-Williams & Challis 2010, 2011:115–18).

Images and ethnography together thus suggest that rock shelters where the San painted were thought of as places of breakthrough between realms: the walls of rock shelters were 'veils' behind which lay the spirit world. Painters could bring the Mantis's 'droves' of eland and rain-animals through that interface. Conversely, it seems likely, although there is no direct evidence, that people could pass through it into the spirit world where they could fight off marauding spirits, beg the Mantis to spare the sick, and learn about spiritual things.[35] The rock face was not a meaningless *tabula rasa* on which image-makers could place anything they wished. On the contrary, it was a distinct context—a mediator between realms.

At a more inclusive level, the Maloti-Drakensberg mountains themselves, of which the rock shelters were an integral part, were an element in an overall conceptual landscape. Qing spoke of 'a place enclosed with hills and precipices, and there was *one pass*, and it was constantly filled with a freezingly cold mist, so that none could pass through it' (Orpen 1874:7–8; emphasis added). The kloofs with their rock shelters, the soaring basalt peaks, and the high, isolated valleys with their mists were together a meaningful setting for myths and images (Lewis-Williams 2010; Lewis-Williams & Challis 2011). Passing through barriers of different kinds, be they water surfaces, rock faces or mists, is a major component—one could say, a foundation—of San religious experience and cosmological thought.

There is thus a conceptual equivalence between the places of creation and growth specified in the myths (waterholes and kloofs) and the locations of painted images of eland (rock shelters). In both myths and imagery, the eland emerges, by an act of creation, from a hidden or invisible spiritual dimension into the level of daily life. Together, images and rock shelters make statements about people, spiritual beings, and animals and their various places in the cosmos and thus about social relations, not just in the spirit realm but also in the material realm. An implication of this understanding is that images transported from rock shelters to the pages of books leave half their meaning behind.

Constituents

MYTHS In the /Xam creation myths, the Mantis places a shoe in the water-hole (Bleek 1924:2, 5). As we saw in Chapter 3, the shoe was probably made from a piece of eland hide, the San's preferred raw material for making shoes.[36] As I have also mentioned, the Mantis fed the eland honey and rubbed it on the animal's flanks as it grew.

The Maloti creation narrative does not mention honey, but it does speak of the role that fat, the food closely related to honey, played in the creation of the eland. After the Mantis's sons have killed the first eland without his permission, he tries to reconstitute it by ordering one of his sons to place some of its blood in a pot and to stir it by twirling a stick between the palms of his hands. The attempt fails, producing 'frightful' snakes and hartebeests. The Mantis then tells his wife to add fat from the eland's heart and to churn the mixture. At first this produces aggressive bull elands, then eland cows, and finally 'multitudes of elands, and the earth was covered with them' (Orpen 1874:4). Parts of an eland (its fat and blood) become the new eland.

IMAGES Although the evidence for the constituents of San paint is slight, there are nevertheless parallels between the eland creation myths and the making of rock art images.

We have no record of honey being used in the making of paint, and experiments have suggested that it is impractical (Johnson, Rabinowitz, & Sieff 1959:16). The use of a part of an eland in paint-making was, however, recorded. In 1930, Marion How asked Mapote, a seventy-four-year-old Basotho man living in southern Lesotho, to demonstrate the practice of painting. As a young person, he had learned, together with his 'half-Bush stepbrothers', to paint with San people (How 1962:33). Given his age, this must have been at or before the time when Qing spoke to Orpen. Today, some of his paintings are preserved in the Origins Centre at the University of the Witwatersrand, Johannesburg. In a now famous phrase to which I return in the next chapter, Mapote said he would paint an eland because 'the Bushmen of that part of the country were of the eland' (Chapter 8).

Among other things, he asked for 'the blood of a freshly killed eland' (How 1962:37–38). No eland being available, he made do with ox blood. But it is clear that eland paintings were *ideally* made from part of an eland—its blood. As How remarks, Mapote's request for *fresh* blood implies that painting took place 'immediately after a successful hunt and a good meal', the sort of occasion on which the Kalahari San almost invariably conduct a trance dance. Bearing in mind what we know about the relationship between eland and potency, we may conclude that dancing and painting were probably, if not invariably, related activities. It will also be recalled that the Ju/'hoan avoidance word for eland was 'dance'.

How describes how Mapote started 'from the animal's chest' and moved 'his brush along smoothly without the slightest hesitation'. He made his brushes from 'bird feathers stuck into the ends of tiny reeds' (How 1962:33). He started to paint at the dewlap of the eland, a curious practice that How noted. It may have been significant, even if aged Mapote himself was not conscious of it: for the Kalahari San, the size of an eland's dewlap is important, because it indicates not only the sex of the animal (the male dewlap is larger than the female's) but also the quantity of its potent fat. Some eland images have excessively enlarged dewlaps.

The use of eland blood in the making of paintings was confirmed in the 1980s by an old woman of mixed San descent whose father had been a painter in the southern Drakensberg (Jolly 1986; Lewis-Williams 1986; see also Lewis-Williams 1995). She also specified fat from an eland's stomach (cf. Kannemeyer 1890; Wells 1933). The woman danced in her father's rock shelter, as, she said, her ancestors had done. When they needed more potency they turned to the images, and the desired power radiated from them and into the dancers. Again, a relationship between painting and trance dancing is implied. The paintings of eland, like actual eland antelope, became, in locations of cosmological transition (walls of rock shelters), reservoirs of potency.

Pigment was also a significant ingredient. Mapote indentified a particular pigment that he called *qhang qhang*; it 'glistened and sparkled whereas the ochre [obtained from a local store] was just a dull earthy substance' (How 1962:34; Lewis-Williams & Pearce 2004a:101). This special pigment 'was dug out of the basalt mountains'. It thus came from the heights of the mountains where the kloofs were 'constantly filled with a freezingly cold mist, so that none could pass through it'. *Qhang qhang* was the only pigment that was mixed with blood. The combination of the two special substances for the painting of an eland was clearly meaningful.

As the first eland was made out of part of an eland (a shoe), so at least some painted eland were also made out of parts of an eland (blood and, probably, fat). There was a continuity of being and potency.

Creators

MYTHS As we have seen, in both major nineteenth-century sources, the eland was created by the Mantis, the trickster-deity; and the Mantis, the creator of the first eland, was also the *ur*-shaman (Lewis-Williams 1996, 1997).

IMAGES There has been some speculation as to which San people made the paintings. The depiction of features that can be seen only by people in trance—the spirit leaving the top of a shaman's head, expelled sickness, transformation into animals, rain-animals, so-called 'threads of light' that weave in and out of the rock face, and so forth—all suggest that at least some of the image-makers were depicting things that they themselves 'saw' and experienced rather than things about which they were merely told. Still, it needs to be emphasised that this does not necessarily mean that *all* painters were shamans or vice versa.

There is an additional link that is, I believe, fundamental to an understanding of San myth and rock art. Guenther perceptively underlines the relationship between, as he puts it, the 'trickster and the trance dancer':

> The trickster and the trance dancer are the two central, key figures of Bushman religion (and the pivotal characters of this study).... We will see the trickster figure as the embodiment of the ambiguity that pervades Bushman mythology and cosmology, much the same as the trance dancer embodies this state with respect to ritual. Both are ontologically ambiguous, confounding such basic categories as natural–supernatural, humanity–divinity, human–animal. . . . Both figures are ontologically fluid, ever ready to change who, what, and where they are, through transformation and, in the trancer's case, transcendence by altered states of consciousness. Both figures are equivalents of each other, the trickster active primarily in the sphere of myth and lore, the trance dancer in ritual, and thereby embedded in real-life society. Yet the two figures frequently enter one another's domain, such as at the trance dance, at its climactic moment, when the trancer 'dies' and his spirit takes over and proceeds with the curing task on a mystical plane. At this moment of the trance dance the trancer's spirit may encounter the trickster-god, who is attracted to the curing dance. The trancer and the trickster are alike in other ways as well; both are curers who may employ identical methods, and in one of his numerous personas the trickster is 'the magician'. (Guenther 1999:4)

This understanding helps to explain what /Han≠kass'o meant when he said that //Kabbo 'had [*/ki*] Mantises, he was a Mantis's man [*/kaggen-ka !kwi*]' (Bleek 1936:143). He was an element in a network of interrelations of which the Mantis was a major node and which included living and dead shamans. This network was to some degree in the minds of all San makers and viewers of rock art and must have informed their 'readings' of painted images. It should be borne in mind as researchers formulate their own understandings.

To these thoughts I add a curious, but perhaps confirmatory, entry in one of Dorothea Bleek's notebooks. In 1911 she undertook a journey to see, at least in part, if she could find any descendants of her father's and aunt's /Xam people. She was largely unsuccessful. She did, however, see rock engravings. Then, on page 188 of notebook A3 004, after an account of digging graves that San descendants had given her, she wrote: '/Kaggen makes the pictures'.[37] Although I have found no confirmation in the Bleek and Lloyd Archive, the note seems to suggest that there was probably a tradition among the /Xam descendants that the Mantis had made the local rock art images. Bleek could hardly have fabricated the note and then never have referred to it again. This possibility—I believe probability—throws new light on Guenther's remarks on the trickster and trancers that I quoted above: 'Both figures are equivalents of each other'. The Mantis may well have been believed to be the maker of the eland antelope *and* of rock art images. Today many people believe that God provided the Bible, although human beings actually wrote it down. Some similar situation may have obtained among the San. Further, the present tense of Bleek's statement suggests that as late as 1911 people believed images were still being made supernaturally.

In sum, we may say that the making of an eland image by a San person was in some ways, consciously or unconsciously, a reiteration of the original mythical creation of the most potent of all the Mantis's creatures. The shaman-painter was, as it were, 'standing in for' the Mantis.

Principles and Nuggets

I have argued that image-making was a closely integrated part of a network of myth and ritual. The parallels between myths concerning the creation of the first eland and the rituals of image-making point to the existence of an overarching framework of belief. The images were parts of the web of San thought and gave indigenous meaning to what are, to Westerners, often inexplicable tales. As Clifford Geertz memorably argues, 'man is an animal suspended in webs of significance he himself has spun' (1973:5).

It seems inescapable that many San must have been fully aware of the significances of verbal nuggets in the myths *and* of the locations and constituents of their painting activities. Instead of formulating a structure of belief, as anthropologists might do, they were probably 'suspended' in the network of significances that I have outlined: they lived within that cognitive framework. The practical working out of existential principles in all the vicissitudes of daily life rather than the construction of abstract 'meanings' was what mattered. That is why I have misgivings about suggestions that certain images symbolised abstract qualities and concepts for the San.

I do not argue that every image derived from a single trancer's visionary experience, although many seem to have done so. In some instances the making of an image (for example, one of the many potency-filled eland) may have prepared for, or accompanied, a dance or even a more solitary transcosmological experience. In some circumstances, the result of making an image may, to a certain extent, have:

- paralleled the outcome of a full trance dance in that it manifested the abilities of shamans to dissipate threats and tensions,
- provided people with valued glimpses of the spirit realm, and
- even performed a prophylactic function by deflecting 'arrows of sickness'.

The making of new images may have been influenced at least in part by what had already been painted on the shelter walls (Lewis-Williams 1972; 1974; 1992b). Painters must have noted their presence. Performances of myths were ephemeral; images were not. There is no evidence that people attempted to erase older images. New images, although informed by existing images and shared memories of trance experiences, probably added to a community's tradition. Each of the dozens of eland images in a shelter was probably individually valued *and treated as if it were unique.*

Novel images (for example, a depiction of crabs) were, however, sometimes injected into this tradition (Dowson 1988; Lewis-Williams & Pearce 2012a). As Biesele says:

> Though individual accounts may be heeded with great attention, there is also a sense in which oral tradition 'swallows up' (in Jack Goody's words)

> the achievements of individuals into cultural anonymity. This observation is true of both trancers' accounts among the Ju/'hoansi and contributions to the folktale tradition. Both kinds of narration embody a kind of timeless, anonymous truth at the same time as they are being creatively renewed by individuals. (1993:76; Goody 1977)

Although writing of /Xam myth, Guenther aptly describes a central feature of this 'anonymous truth' as it appears in San rock art:

> The transition between the real and surreal realms is seamless, and the /Xam storyteller . . . appears to give no indication where in the narrative the dream left off and reality set in. As a consequence, the plot of such narratives can become dense and incoherent to a reader accustomed to more linear and unequivocal perceptions of reality. (Guenther 1999:104)

This passage aptly sums up the kind of fundamental, framing beliefs that San viewers of rock art probably had in their minds when they viewed and related to 'dense and incoherent' panels of images. The logic of San myth tells auditors that there is more than one way of regarding the cosmos and one's place in it. The painted images, mixing realistic depictions with bizarre conceptions and referring in various ways to what Guenther calls 'surreal realms', showed viewers that the rules of mundane material life can be suspended: transition between realms was 'seamless'. Contradictions and ambiguities are not merely acceptable but rather the essence of life.

CHAPTER EIGHT

People of the Eland

The Mantis is with the eland as it lies dying.
(Diä!kwain, L.V.18.5359')

To leave the matter of integrated San thought with the suggestion that narrating and painting were in some instances parallel activities would be to stop short of investigating the possibility of links between, on the one hand, eland myths and images and, on the other, the San people themselves in their daily lives. Most rituals and myths in preliterate societies are performed in social circumstances; they respond to—and act on—the people involved. In this chapter I therefore ask: how did social pressures created in the southeastern mountains by adjacent Bantu-speaking agricultural people and, more recently, encroaching colonists affect the San? Did the resulting land-losses, conflicts, intermarriage, and rainmaking services remould the San's own sense of identity, as well as the other people's concepts of them? Were the performances of myths and the painters' reenactments of the Mantis's creative act stimulated by a desire, perhaps not fully articulated or rationalised by every San person, to sustain the identity of the San as a distinct people with their own economic base and religious system in the face of increasingly powerful and often hostile communities?

As we have repeatedly seen, San thought seems never to comprise simple, one-to-one relationships of the kind that Western researchers mistakenly tend to seek. Rather, the creation of eland, either originally by the Mantis or subsequently by image-makers, was part of a larger pattern of significances. To explore that pattern further I return to Marion How's 1930 interview with Mapote (Chapter 7). In some ways it was a ghost of Orpen's 1870s' meeting with Qing. In both cases, the Westerners had little or no understanding of

Myth and Meaning: San-Bushman Folklore in Global Context by J. D. Lewis-Williams, 173–181

San life and beliefs beyond colonial stereotypes. But they were nevertheless able to record information that we can today evaluate in the richer San context that we now have. We can spot significances that passed them by. For instance, as we saw in Chapter 7, How recorded evidence for the ingredients of San paint, ingredients that we can now see fitted in with San notions of the recreation of an animal from parts of it.

Neither Orpen nor How spoke a San language. Unlike the Bleek family, they both depended on interpreters. Therein lies a problem: we do not have access to Mapote's original words. In cases like this, we are, however, sometimes able to discern a San idiom in its, probably inadequate, English translation. In this chapter I deal with one such instance. Speaking as a Basotho man who had known the San all his life and had, as a child, learned to paint with them in their rock shelters, Mapote said that 'he would paint an eland, as the Bushmen of that part of the country were of the eland' (Colour Plate 9; How 1962:38). His phrase was popularised in the catchy title of Patricia Vinnicombe's 1976 book, *People of the Eland*. Like the other nuggets I have tried to unpack, Mapote's statement is not as vague as it seems in English translation (Lewis-Williams 1988).

An Elusive Preposition

At first glance it may appear that How's use of the preposition—'*of* the eland'—is simply an approximate translation and, perhaps, condensation of a meaning that is now lost forever. I argue that it more probably reflects an idiom that is preserved in the Bleek and Lloyd Archive. We can now see that Mapote's phrase dovetails with the interrelated San beliefs that I have so far discussed.

In previous chapters we saw that the /Xam spoke of *!khwa-ka !gi:ten*, a phrase that may be translated as 'shamans of the rain'. *!Gi:ten*, it will be recalled, is the plural of *!gi:xa*, one who is 'full of' *!gi:* ('potency'); the suffix *–ka*, sometimes *–ta* or *–ga*, forms the possessive (Bleek 1928/29:93). *!Khwa*, as we have also seen, means rain and water. Occasionally, a speaker used the phrase *!khwa-ka !kwi*, literally, 'a *person* of the rain' (Bleek 1933:306). *!Kwi* means simply man or person, but the /Xam understood the phrase as a whole to mean a '*shaman* of the rain'. This is an instance of what Biesele (1975:176) describes as San elision: 'The beginning of a phrase may be uttered and broken off, but the entire phrase must nevertheless be inferred from context'. She was writing of San songs (Chapter 9), but the principle applies in ordinary speech as well: in some contexts, it was prudent to say *!khwa-ka !kwi* rather than *!khwa-ka !gi:xa*, especially if the *!gi:xa* was a spirit (Chapter 9).

The /Xam spoke of shamans who had associations with various other things, not just rain. They included springbok, locusts, and mantises. For instance, a female shaman, Tãnõ !khauken (*!khauken*, 'to tremble as in trance or illness'), who was associated with springbok, was said to be a *wai-ta !gi:xa* —'shaman of springbok' (*wai* means 'springbok'; Bleek 1935a:43–47; cf. Bleek 1936:142–43). Tãnõ !khauken and her springbok provide an illuminating vignette of San life that helps us to understand San spirituality as it was manifest in daily life. She said that she had a 'short-horned springbok' and that her father had inadvertently killed it. She explained that it was her son's springbok, that it was castrated, and that she normally kept it 'tied up'. It was not a 'food springbok'; it did not 'wander about' but was kept on a 'thong' (*!hãũ*). (The /Xam word here is the one used to mean the mystical thong that shamans of the rain employed to capture a rain-animal.) Loosed among wild springbok, Tãnõ !khauken's springbok led the herd toward the /Xam camp and the hunters.

How should we understand Tãnõ !khauken's statements? Real springbok are too intractable to be kept tied up, let alone trained to act as a decoy, so it seems that she was speaking about a conceptual springbok. Her control of springbok took place in a context of interaction between the material world and the spiritual dimension: the two realms were interwoven. As we have seen, and as in probably all religions, San people believed in a spirit realm that they could visit and also in the presence of elements of that realm in daily life. Rigid casuistry is not characteristic of San thought.

Tãnõ !khauken added an additional point. She thanked her mother for making a 'cap of the springbok's head'. She explained that the springbok would follow the wearer of such a cap. In another instance, a man called !Gurriten-dé also claimed to be a 'springbok sorcerer' (*wai-ta !gi:xa*), that is, a sorcerer *of* springbok. He too made a cap from the scalp of a springbok and sewed it so that the ears stood up. He said that he 'had' (*/ki*) springbok (Bleek 1936:144). Rock paintings of people wearing distinctive eared caps are common in some regions. They are painted nuggets that situate many groups of images in those regions.

In a related usage, */ki* means to possess the supernatural potency of designated animals that San shamans harness to perform their tasks. Some painters expressed this relationship between shamans and their animals in a dramatic, visual way. Asked to comment on George Stow's copy of an Eastern Cape Province group of thirty-one human images, many of which appear to have antelope heads, Diä!kwain did not say they were wearing animal masks for the purpose of stalking, as some rock art researchers have supposed such images to depict (Stow & Bleek 1930:pls. 13 and 14). Another of Stow's copies, one that apparently shows a stalking bowman disguised as an ostrich, is sometimes used to support this hunting interpretation of apparently antelope-headed images (Stow & Bleek 1930:pl. 21). His 'stalking ostrich' copy is, however, almost certainly a fake that he concocted by drawing on illustrations of the supposed San practice in books written by early travellers (Dowson, Tobias, & Lewis-Williams 1992, 1994).

Rather than speaking of 'masks' and hunting, Diä!kwain said that the people depicted were wearing caps and that they 'mean to tread the 'ken with them. At the time when they do the 'ken they wear such caps' (Stow & Bleek 1930:captions to pls. 13 and 14). The figures in the painting are therefore performing a trance dance by activating *//ken*, 'potency'. It should be noted that the images in Stow's copy are among the very few that clearly show some sort of headgear; most such images show the head blending seamlessly with the body in a therianthropic image (Lewis-Williams & Challis 2011; Lewis-Williams & Pearce 2004a).

But why did the /Xam wear eared caps in a trance dance? There seem to be two reasons. First, they wished to blend with the animal from which they derived their potency: '!Gurriten-dé was a springbok sorcerer, he had [*/ki*] springbok' (Bleek 1936:144). As a Ju/'hoan man said: 'When you see the animal in the fire, it changes you. You become that animal' (Keeney 2003:85). Second, they were probably performing their task of guiding antelope in the context of the out-of-body travel that they experienced in a trance dance. Hunting played a role in the dance.

In sum, ethnographic and linguistic evidence shows that Mapote may have been using a common San idiom—'of' an animal —that was used to speak of shamans who were associated with, or 'possessed', and were transformed into eland. How does that understanding fit in with what else he had to say? His words have a wider implication.

A More General Meaning

We find this further implication of Mapote's statement in the Kalahari San ethnography. The Ju/'hoansi do not define themselves as a people by, exclusively, race or their foraging economy. Instead, they sometimes call themselves 'owners of *n/om*' ('potency'). In some contexts, they say they are 'the owners of Giraffe Music' (Marshall & Ritchie 1984:2), that is, giraffe potency. The Giraffe Medicine song is still popular in the Kalahari (Chapter 2). Despite its popularity and the people's use of it as a defining characteristic, individual Ju/'hoansi still insist that it is eland that have more potency than any other creature. As we have seen, the potency of a song or dance is believed to be in the (largely wordless) music. This also appears to have been the case in the nineteenth-century south: Qing said that 'Cagn gave us the *song* of this dance, and told us to dance it, and people would die from it [that is, enter deep trance; for example, Katz 1982:115], and he would give charms to raise them again' (Orpen 1874:10; emphasis added). The 'song' was a gift from the Mantis, he who created the first eland.

The Ju/'hoan case shows that *all* the people in a San community share in some way in the potency that their shamans possess and activate to care for them. Biesele says that *all* Ju/'hoan spectators at a trance dance participate in the spiritual uplift that derives from contact with the supernatural realm: 'Contact with the beyond is *regularly* made, and *all* who come to the dance experience an uplifting energy which they feel to be *a necessary part of their lives*' (Biesele 1993:74; emphases added). This is an important observation. The panoply of animal potency, although principally manipulated by shamans, protects *everyone* from sickness and danger sent by malevolent spirits. Similarly, all viewers of images probably shared comparable ideas about the benefits inherent in eland.

It is therefore likely that, in his phrase 'the Bushmen of that part of the country were of the eland', Mapote was drawing on a San idiom that he had picked up from his half-San stepbrothers. Although he was not himself San, he used the San phrase to speak of an extension from the principal controllers of potency to all San people known to him. In this sense, the Mantis's mythical creation of the first eland was not only the creation of the foundation of San religion and contact with the supernatural, as I have so far argued. It was also the creation of the San as a distinct people: not just their shamans, *all* of them were 'of the eland'.

Similarly, painting an eland made a multicomponent statement about that antelope's exceptional potency, shamans' possession of eland potency, their ability to control eland by leading them to hunters, their skill in drawing them through the rock face in the form of images, and, ultimately, by themselves becoming eland. *In toto*, the act of painting an eland probably proclaimed the Maloti San as, complexly, 'people of the eland'—people united by the supreme potency of the eland antelope.

Because of interaction with their Basotho neighbours, it is probable that *both* groups of people recognised the San association with eland, as indeed Mapote's statement implies. After all, despite having lived with San people, Mapote was drawing a distinction between them and himself. The Ju/'hoansi sometimes speak of themselves as being 'the owners of Giraffe Music'; the Maloti San were 'people of the eland'. Mapote realised that he was not one of them. Whether San painters consciously intended to manifest the distinction between themselves and other people every time they made an eland image is another matter. I suspect that the primary, conscious reason for painting eland so frequently probably remained the creation of reservoirs of protective potency that could actualise the spirit realm. It was from that intention that wider issues flowed.

Neighbours

The Drakensberg San's neighbours are important in this context. This is the 'contact period' that I discussed in Chapter 2. I now make some more specific points. People construct their own identity *vis-à-vis* other people and *their* perceived identities. I suggest that emphasis on San identity as being 'of the eland' probably increased during the last 2,000 years as Khoekhoe herders, Bantu-speaking agriculturalists, and, subsequently, colonists occupied the land and destroyed the vast herds of antelope, especially eland, the Mantis's favourite creature.

Orpen (1874:3) refers to the 'game country' being occupied by Bantu-speaking farmers. Until that time, the agriculturalists had lived 'on good terms' with the San, who 'succoured them' in times of need. The early traveller John Barrow found that 'even in 1806 . . . the dutch [*sic*] farmers had almost destroyed the species [eland] by their inconsiderate (meaning too frequent) hunts' (Arbousset & Daumas 1846:45). A number of other early writers also commented on the rapid extinction of the comparatively docile eland. By the second half of the eighteenth century the eland was extinct in the western Cape Colony,[1] and by the end of the nineteenth century only a few eland survived in the 'fastnesses of the Drakensberg' (Bryden 1899:422).

In this destructive period, the San themselves suffered greatly, especially at the hands of colonists, who organised punitive raids against them.[2] At the end of the nineteenth century a missionary found that, like the eland, only a few San 'were still to be found, dragging out a precarious existence in the innermost fortresses of the Malutis' (Widdicombe 1891:14). Qing's life was part of this tragic history: as we saw in Chapter 2, he had 'escaped from the extermination of their remnant of a tribe in the Malotis' and entered the employ of a Phuthi chief as a hunter.[3] Other San people joined ethnically mixed bands, which were nevertheless known as 'Bushmen', and continued the San painting tradition.[4] Some acquired small numbers of cattle, sheep, and horses and depicted them.[5] An interesting aspect of the no doubt shifting relationships between the San and the farmers can be gauged from early nineteenth-century reports. The missionaries Thomas Arbousset and François Daumas found that Basotho neighbours of the Maloti San said that the eland herds were tended by 'an imaginary shepherd' who was called Unkonagnana (Nkonyana, in modern orthography), which means 'little nose'. He is 'never seen by human eye' (Arbousset & Daumas 1846:46). There is a strong suggestion here that this notion derives from San beliefs about the Mantis, of whom Qing said: 'Where he is, eland are in droves like cattle' (Orpen 1874:3), and of whom Diä!kwain told Wilhelm Bleek that he could be 'by you' even if you could not see him (B.XXVI.2463 rev.; cf. Hammond-Tooke 1997). The Basotho also believed that eland meat was infected by 'venomous juices' and they had to purify themselves before eating it. Perhaps most interesting, they said that 'a very dangerous yellow viper' hid in the hair between an eland's horns; this tuft of red hair is a prominent feature of many rock paintings of eland. They therefore struck the top of a killed eland's head with a stick (Arbousset & Daumas 1846:46). This Maloti practice echoes the /Xam belief that 'the Mantis sits between the Eland's horns' (Bleek 1924:11). The Mantis, it will also be recalled, could change himself into a snake. At a remarkable site in the Free State province of South Africa there are numerous depictions of detached eland horns and the tuft of hair between them. They are a painter's idiosyncratic emphasis on that part of the antelope (Lewis-Williams & Pearce 2004b).

The antiquity of at least some Maloti San religious beliefs is suggested by archaeological finds. A depiction of the distinctive arms-back San trance dance posture on an excavated painted stone that has been securely dated to 200–350 C.E. shows that the dance was being performed by at least the beginning of the period of contact with Bantu-speaking peoples (Mazel 2009a, 2009b).[6] In addition, research suggests that there was, also at about this time, an increase in the painting of elaborately detailed, shaded polychrome images of eland (Mazel 2009b). This increase probably implies ritual intensification in the face of ecological destruction, political uncertainty, and emotional stress. Referring to the southeastern rock paintings of this period, the archaeologist Aron Mazel notes the central role of the trance dance in San life and thought and suggests that

> a group of potent and innovative shamans may have inspired the emergence of a new painting tradition through translating a series of particularly powerful and remarkable visions onto the rock face in the form of shaded polychrome eland and associated features. (Mazel 2009b:109; cf. Dowson 1994, 2000)

This suggestion is supported by Guenther's experience in the Kalahari (Guenther 1999:196; Pearce 2012). Working with the Nharo San at Ghanzi, he found that the collective activity of the trance dance became a 'suitable vehicle for cultural revitalization' in a period of land-loss and suffering as other peoples occupied their land and destroyed their livelihood. Similarly, Geoffrey Blundell (2004:130) concludes that San rock painting in the Malotis and adjacent areas 'in the mid-nineteenth century, like the Great Dance at Ghanzi during the mid-1970s, became a vehicle for the expression of San identity'. The evidence thus suggests that 'the eland may have formed a vital part of the hunter-gatherer spiritual toolkit in dealing with the substantial and irrevocable changes that were undermining the world that they had known since the beginning of time' (Mazel 2009b:110).

As their society and way of life came under threat, San shamans stepped up their protective rituals and thereby emphasised their connection with, and dependence on, the albeit dwindling eland herds. Some took the initiative and developed interethnic relations by becoming rainmakers for Bantu-speaking agriculturalists.[7] The material rewards that they received for this service probably exacerbated tensions within San communities. Some San rainmakers began to possess small numbers of cattle that they obtained as rewards from their Bantu-speaking neighbours. In this way, the social inequality that had been created by possession of ritual knowledge and experience acquired a material dimension. Traditional spiritual techniques were being harnessed in new economic circumstances.

This nexus of rainmaking and relations with other, sometimes hostile, communities brings us back to the /Xam song with which this book began:

it concerns the killing of the rainmaker !Nuin-/kuiten, as we shall see, a man with an interesting name. We are now in a better position to answer the questions that I posed at the end of Chapter 1:

- Who are the 'people' mentioned in the song?
- What is the 'string' that they broke?
- Why did they break it?
- In what sense was the place thereafter not 'pleasant'?

CHAPTER NINE
The Broken String

For, he had desired to take my father with him,
that he might teach father his magic which he worked.
(Diä!kwain, L.V.15.5098–99)

Now that we have some idea of the complexity and interrelatedness of San thought as it runs through myths, rituals, and rock art images we can return to Diä!kwain's 'Song of the Broken String'. Clearly, it is not comparable to an isolated piece of Western 'poetry' composed by a 'poet'. On the contrary, it came out of the rich and turbulent San milieu we have been exploring and can be understood only within that milieu. We must therefore look more carefully at the various meanings of a simple yet elusive word in the song: it is 'people'. The Maloti San, we saw, were 'people of the eland'. Who were the 'people' who broke the string?

As far as the general public today is concerned, 'The Broken String' is arguably the most famous text in the Bleek and Lloyd Archive.[1] Lucy Lloyd herself must have considered the song important (although exactly what she thought it showed about the San is not clear), because she extracted it from its narrative context and published it separately in 1911 with her own lineation in a section of *Specimens of Bushman Folklore*. She entitled the section 'Poetry' and so cast it in a context of expectation foreign to its San origin (Bleek & Lloyd 1911:237). Later, Dorothea Bleek published a version, also with poetic lineation, as (more appropriately) part of a longer text (Bleek 1936:134; see Prologue).

As the song stands in the published versions, it strikes a chord in many present-day readers, and they mould it to fit their own deeply held feelings about the destruction of the southern San as a network of foraging communities and indeed about the extinction of the /Xam language itself. When

---

Dorothea Bleek visited the /Xam homeland in 1910 and 1911, she 'found just a handful of old people left here and there, some of them relatives of our former men' (Bleek 1924: unnumbered page). By the early decades of the twentieth century the traditional /Xam way of life existed only in a few diverse memories and in the pages of the Bleek and Lloyd Archive.

From this perspective, it is understandably easy to interpret the song as the breaking of a 'string' (however conceived) that connected, first, Diä!kwain to his earlier life; second, Xa-ttin (Diä!kwain's father) to the rainmaker !Nuin-/kúiten; and, third, us to the San's pristine past as it was before it was destroyed by the advancing frontier. I do not denigrate these valedictory readings: the words of the song have come to be a monument to a departed way of life and a southern African genocide. Instead, I try to add to that view by seeking a more indigenous and contemporary understanding, in fact to explore a palimpsest of meanings.

The context of the song in Lloyd's notebook is key to forming an indigenous understanding of the words of the song. Lloyd began to note the text down on July 26, 1875. Diä!kwain's overall train of thought is, however, important. It seems to have started even earlier, on July 22 in a passage about 'sorcerers'. Lloyd entitled it 'How Sorcerers Assume the Form of a Jackal, or of a Bird, and How They Then Act, etc.' (L.V.14.5055–78; Bleek 1935a:15–19). I therefore go deeper into the activities of San shamans than we have so far ventured. It is here that we find insights into what was in Diä!kwain's mind when he gave Lloyd the song. We have seen that *!gi:ten* were a bridge between existential realms, that they healed the sick, made rain and guided antelope. But we are left wondering what it was like to live daily in a hunter-gatherer community in which certain people were believed to have especially direct contact with the supernatural realm.

'A Man Who Is a Sorcerer'

Diä!kwain's train of thought began with an event that was common in /Xam life: people travelling between widely separated camps. Sometimes the journey was long and could take a couple of days. The rough map that Wilhelm Bleek drew suggests journeys of as much as fifteen days (Chapter 1). The people then carried ostrich eggshells of water with them to sustain them in the parched terrain (Lewis-Williams 2002a:52–77). Always, there was the threat of wild animals that could snatch them at night. Lorna Marshall vividly describes a cameo of Kalahari Ju/'hoan life and the way in which realms intermingle seamlessly:

> A little group of two families was making a long journey. They were encamped in the night alone in a vast, flat space. A lion came and prowled around them. The moonlight was bright and the people could clearly see the lion circling. They were terrified.... One of the men, Bo, cried out, to the lion, 'You lazy beast! Why do you not go and kill an animal instead of coming after us? We are not your equals'. The lion growled and did not go away. Toward morning, one supposes in response to the long-continued emotional stress, Bo fell into a trance. At sunrise the lion left, and the people said that Bo's spirit followed it and chased it far away, and they never saw it again. When Bo's spirit returned to his body and he came out of trance, his nose bled severely. (Marshall 1969:374)

That was in the 1950s, but the same sort of thing must have been experienced by the nineteenth-century /Xam. It was natural that people would worry about friends and relatives who embarked on a long journey. As Marshall's account also shows, this was one of the circumstances in which shamans played a role in daily life.

When Diä!kwain described people setting off across the plains to join relatives at another camp, he spoke of 'we', but it seems that he was probably not speaking of a specific instance.[2] He was rather telling Lloyd about what benign San shamans could be generally expected to do in such circumstances. As always, his exact words are important.

After the departing people have left, a 'sorcerer' 'turns himself into a jackal' because he wants 'to go about taking care of us'. The travellers do not see him 'when the sun is high', but when the sun sets 'then we see him in the evening. ... Then it is that he becomes visible'. But not in his human form. While they are resting after their long walk 'which has made our legs ache ... he sits barking like a jackal.... [T]hen we know that he is asking us whether we are still as well as we were when we left him' (Bleek 1935a:16). The people assure him that they are safe and well. 'As soon as he sees that we are really among our people, he returns to his home'. Meanwhile, at his home 'the people still see his body, in which he is a man, that is at home' (Bleek 1935a:17). His human body has remained visible at home during his absence in the form of a jackal.

As Lloyd's title indicates, other /Xam shamans could turn themselves into a 'little bird ... that flies about our heads' (Bleek 1935a:18–19). Sometimes, the bird 'sits on our heads . . . he sits peeping at us to see if we are still as we were when we left him'. The people respond to the bird: 'If we are wise people . . . we talk to him and tell him, that nothing has happened to us'. Satisfied that the people are safe, 'he flies away, he chirps, just as a little bird does when it flies away'. One of the people says: 'I knew it was not a little bird, but it was you. I knew that you were the one who had come to see me'. For the /Xam, the natural world that they knew so well was, at least from a Western point of view, ambiguous: shamans could be animals and vice versa, but this ambiguity posed no problems for the San.

All this talk of shamans and their transformations, we should not forget, was new and strange for Lloyd. The intermingling of the natural world with supernatural transformations in such mundane circumstances as an inter-camp march and to such a marked degree as transformation into an animal was unknown to her. It is therefore not surprising that she sought further explanation from Diä!kwain. The first of her explanatory notes on a verso page tells us what the people remaining at the shaman's camp saw: 'D. H. says that the sorcerer, who turns himself into an animal, returns at cockcrow, before day break, while the people are still sleeping, & do not see him come. His human form remains meanwhile sleeping at home' (L.V.15.5068').

Lloyd headed a second and longer note: 'How Xa-ttin (D. H.'s father) asked a dead sorcerer named !Nuin-/kúi-ten for rain, w[h] was speedily bestowed' (5068'; Bleek 1933:382). This is the first mention of !Nuin-/kúiten, the man whose death gave rise to 'The Song of the Broken String'. The people we have so far considered are 'ordinary' or 'general' shamans. By contrast, !Nuin/kúiten was a rainmaker —a *!khwa-ka !gi:xa*.

!Nuin/kúiten

When Lloyd began the notebook partially reproduced in the Prologue to this book, she immediately took up the intriguing ideas of which she had been hearing. She headed the first entry: '!Nuin-kúiten (who was a sorcerer or magician)' (L.V.15.5079). Lloyd's parenthesis shows that she was struggling to find an appropriate English word for *!gi:xa* (Chapter 2).

Another note (L.V.15.5078'; see also 5084ff) casts light on a characteristic of San 'sorcerers', one that applies especially to !Nuin/kúiten. It is fundamental to an understanding of 'The Song of the Broken String'. In addition to jackals and birds, San shamans could, commonly, turn into lions, as we saw in

Chapter 5. Diä!kwain's father, Xa-ttin, told him explicitly that '!Nuin/kúiten used to become a lion'. /Xam people recognised !Nuin-/kúiten's footprints when he was in the form of a lion: 'he walked treading upon hair . . . the hair of the lion's own feet'. Today the Kalahari San still point to the small prick marks in the sand that are made by the hair around a lion's paw. Indeed, one of the /Xam respect words for a lion was 'hair' (*/kuken*); another was 'lighting in' (*/kerre-/e:*), the exact meaning of which is unclear.[3] If children utter the usual word for lion (*//khã:*), the lion may consider it an insult and attack them (Bleek 1932:53).

Another of Lloyd's notes gives more about !Nuin-/kúiten: 'A sorcerer who was, D. H. says, a friend of D. H's paternal grandfather, & was seen by D. H's father when the latter (xã-ttin) was a young man. He taught xã-ttin about "Khwa ka tiken-tiken"' (L.VIII.15.facing 5079). This note suggests that Lloyd at first misunderstood !Nuin-/kúiten's relationship to Diä!kwain's father: earlier, she had written: 'My [i.e., Diä!kwain's] grandfather was !Nuin-/kúiten' (5068'). The note shows that !Nuin-/kúiten was a rainmaker and that he passed his skill on to Xa-ttin, Diä!kwain's father. He taught Xa-ttin about 'the doings of rain', that is, at least in part, the singing of rainmaking songs. The word *tiken-tiken* (a reduplicative plural), however, adds a new dimension of which Lloyd was apparently unaware. In addition to 'things', and unless there is a significant tonal distinction, *tiken* can mean 'leg' (Bleek 1956:203). In this context, the word may be, I suggest, a reference to 'the rain's legs', as the /Xam called the columns of rain that can be seen catching the light as they fall from a distant thunder cloud.[4] It will be recalled that one way in which the /Xam thought of the rain was as an animal that shamans of the rain could capture in a waterhole, lead across the country, and then kill where rain was desired (Chapter 3).

Some *!gi:ten*, like !Nuin-/kúiten, continued to exercise their powers after death and people prayed to them for rain. For instance, Diä!kwain said: 'Mother used to tell me that the spirit people [*/nu:-!k?e*] were those who had been game sorcerers [*ʘpwaiten-ka !giten*]. When they died their thoughts, with which they had been sorcerers and worked magic [*!gi:-ka didi*] continued, though they died and we did not see them' (Bleek 1935:35).

The 'spirit people' had formerly been living 'game sorcerers'. /Han≠kass'o made a comparable point: 'Dead people [*/nu/nuken !ke:*e] who come out of the ground are those of whom my parents used to say, that they rode the rain, because the things [*!hōũ-!hōũ*] with which they held it were like the horse's reins' (Bleek 1933:305; see also Bleek 1933:304; 1935a:24–26). Here we find that deceased shamans of the rain controlled the movements of a rain-animal by means of a thong ('reins'). This is clearly important for our understanding of the song and its 'string'.

There is another intriguing connection between !Nuin-/kúiten and rain. In a tale about a girl at puberty we learn that the girl, taken up by a whirlwind, is transformed into a large snake. The name of this snake is given as //Kheten. But it has another name too: !Nuin—an additional link between the man and rain. When the transformation takes place, 'the folk who possessed their noses [*/nũ*]' sang: 'Ye are now those who behold that that maiden, she now yonder ascends the sky, the rain is now the one who takes her away; she becomes a snake' (L.V.13.5020–21').[5] No doubt, as a result of discussion with Diä!kwain, Lloyd inserted an explanatory parenthesis after 'noses': '(sorcerers)'. The nose, from which trance blood fell, was thought to be the seat of a /Xam shaman's potency.[6] */Nũ* could also mean a spirit and a dead shaman (Bleek 1956:350). The singing of the shamans probably means the activation of potency, as singing does at a trance dance and as it probably means in the tale of the bag in a hole in the ground (Chapter 6). Shamans of the rain were thus believed to be in some, for us not fully understood, way involved in the transformation of a girl at puberty. What is clear is that, for the San, the fundamentals of shamanism permeated, in general terms, other, sometimes apparently quite different, areas of /Xam ritual and belief.

Rainmakers in Daily Life

This division of rainmakers into living and dead points to a theme that we have encountered a number of times in previous chapters: conflict between shamans (and groups of shamans) in the spirit realm. This sort of conflict seems to have much concerned the /Xam. Distant thunder, for example, was spoken of as shamans fighting one another. /Han≠kass'o said: 'When the clouds were thick, . . . then, the clouds were lightening, on account of it. And my grandmothers used to say: 'It is *≠kágára*, with *!hãũ*' (Bleek & Lloyd 1911:119). ≠Kágára and !Hãũ were dead or spiritual shamans of the rain. Natural phenomena were routinely spoken of in terms of conflicting spiritual beings and interaction between realms.

The reality of living *!gi:ten* in ordinary /Xam daily life is seen in an account that //Kabbo gave (Bleek 1933:306–12). It is another glimpse of real-life tensions as they were lived out in a /Xam camp. On the central southern African plateau rain frequently comes in the form of what the /Xam called a 'male rain-animal'—a highly localised thunderstorm. As a result, an area may remain parched while another, a comparatively short distance away, may receive rain. Plant foods and sweet new grass soon spring up there and antelope are attracted to the area. The /Xam would then move camp across the semi-arid terrain to the renewed area. In addition, if people found that the water in their waterhole had turned bitter, a man could ask 'his relation' who was 'a rain man' (*!khwa:-ka !kwi*, 'rain's man') to make rain for the people. If rain fell, they would be able to 'travel away from the old hut and travel to the new water which is sweet, for we have been drinking bitterness' (Bleek 1933:307).

In the instance of which //Kabbo spoke, the rainmaker lived at another camp some distance away. A man's wife agreed that her husband should go to the rainmaker: 'Go and talk to the old man, that he may make rain fall for us'. The man said he would do that and added: 'My younger brother will go with me; we will go and ask that old man that rain may fall'. The rain would then fall 'bringing us back the springbok'.

There was another side to rainmaking. In times of drought, being a rain-controller could attract people's wrath if their rainmaking failed. Diä!kwain spoke of //Kunn, who was blamed for a lack of rain (Bleek 1933:385). The people were not always appreciative of his efforts: 'Then when rain fell, they did not remember that they had asked him to make it fall' (Bleek 1933:386). Like people in any other society, the /Xam had their quarrels: some people were ungrateful, some short tempered, some perhaps vain.

Another shaman was /Kãũnũ, who was murdered (Chapter 7). /Han≠kass'o said: 'A real medicine man he was.... He is now dead... /kãũnũ used to strike the bow-string, and then the clouds came up while we were asleep.... We were asleep; he sat and took up the bow, while we were asleep' (Bleek 1933:390–91).

/Kãũnũ was a real-life rainmaker, not a spirit. He was also able to control the flight of locusts (Bleek 1936:9). //Kabbo said that /Kãũnũ's other names were !Khwa:-ka /kãũnũ and /Kãũnũ !kwa: 'Rain's new grass' and 'New grass leg' (on San names see Lewis-Williams 2013b). Playing a musical bow was one of the techniques that shamans of the rain employed. The practice raises the question of !Nuin-/kúiten's name.

What Does !Nuin-/kúiten's Name Mean?

San names are often biographical. The name given to a person at birth is frequently extended with a nickname. Some are analysable, although problems do arise because of the way in which Bleek and Lloyd noted down what they heard (Lewis-Williams 2013b). They recorded /Xam words as accurately as they could each time they heard them. As a result, there are no definitive spellings. The *Bushman Dictionary* (Bleek 1956) often gives alternative phonetic spellings of words. None the less, the various means by which Bleek and Lloyd obtained the meanings of /Xam words in the first place lend confidence to their work.[7] As a preliminary example we may take the name Tānō-!khauken (Chapter 8). This woman, Diä!kwain's aunt, was a *!gi:xa*: she could turn herself into a lioness (Bleek 1935a:43). As we have seen, the second part of her name, *!khauken*, means 'to tremble', as a shaman does. In the light of this translation of *!khauken*, the first part of the name may derive from *taŋ*, 'to feel, be in pain' (Bleek 1956:191; the tilde and *ŋ* both indicate nasalisation). Today the Kalahari San frequently speak of the pain that shamans endure. Both parts of the woman's name were thus probably related to her position as a shaman.

The shaman !Nuin-/kúiten's name is similarly, though also tentatively, explicable. The first part, !Nuin, may refer to a kaross (*!nuiŋ*; Bleek 1956:484). As we have seen, a /Xam shaman was said to be reluctant to lay down his kaross, even if it is hot, because 'his inside is cold' (Bleek 1935a:13). But, given the context in which we encounter this name, !Nuin more likely means 'sinew, ligament, bowstring, thread' (Bleek 1956:484). Because he was a 'sorcerer' who was associated with a 'string', this translation may be more appropriate than 'kaross'. Again given the context of the name, the second part, */kúï*, may mean 'to hear' (Bleek 1956:324). It is given in the *Dictionary* without diacritics. All these connotations and possible meanings may come together in the /Xam word */kwi-ta*, which the *Dictionary* gives as 'stretch out', as in 'he arises, he stretches out his thighs' (Bleek 1956:334).[8] This translation of the two parts of the name !Nuin-/kúiten as 'stretched out (bow) string' may not seem fully convincing, but other possible meanings of the words as recorded in the *Dictionary* do not seem to fit the context as well as this one. Either way, the whole passage and especially the song itself imply he was associated with and spoke about the sound and sensation of a vibrating and breaking string.

A supplementary but important point emerges from the manuscript. Lloyd at first translated *!nuin* as 'thong' and later changed all three instances to 'string' (see facsimile). I suggest that when she heard the song for the first time, she already knew, from previous discussions, it was going to be about rainmaking. She was therefore thinking in terms of the thong (usually *!hãũ*) that she had previously heard was used to lead out the rain-animal. Then, in discussion after the text had been taken down, the speaker must have denied this reading and insisted on 'string' because he was talking about a musical bow that was used in some rainmaking techniques—as indeed befits the man's name—not about the thong used for capturing a rain-animal. The breaking of his bowstring spoiled his rainmaking abilities and so the place became 'barren'.

The ethnography that I have so far cited sets us up for an informed approach to the 'Song of the Broken String'. The text preceding the song itself indicates that !Nuin-/kúiten was a man who, after being shot by a Boer whose ox he had killed, passed his rainmaking techniques and songs on to Diä!kwain's father. Diä!kwain's father then composed another song, 'The Broken String'. It does not seem to have been a rainmaking song in itself. Biesele (1975a) writes of two kinds of Ju/'hoan songs: those that are potency filled and those that are sung purely for the pleasure or personal concern of the singer.

The Killing

!Nuin-/kúiten 'walked about' in the form of a lion because he wished to find 'those who wanted to kill people' (L.V.15.5087–88). Here Diä!kwain is speaking about malevolent spirits of the dead or other shamans who, sometimes in the form of a lion, dragged people from their dwellings. Real lions, too, were a danger for the /Xam. For example, ≠Kasin, another /Xam man, spoke of a lion leaping into a camp and carrying off his brother (L.V.1.3453–58). In the course of his narrative, Diä!kwain refers to the ambivalent character of at least some, perhaps all, /Xam *!gi:ten*. !Nuin-/kúiten admitted that he could be easily provoked to anger, and he feared that he might kill a person when so provoked.

It was when he was in the form a dangerous lion that !Nuin-/kúiten killed a Boer's ox. Then 'the boer called together a commando against him, the boer went to shoot at him' (L.V.15.5090). Wounded, he staggered 'slowly (and) painfully' (L.V.15.5095) home and told Diä!kwain's father about it. At first Xa-ttin doubted !Nuin-/kúiten had been mortally wounded, but he soon realised that it was indeed true.

The dying !Nuin-/kúiten wanted to take Diä!kwain's father with him into the spirit realm so that

> he might teach father his magic [*ddi-ddi*, 'doings']. . . . Then father would know the things he had taught him (L.V.15.5099). . . . At another time father must sing the songs which he had taught him, father must sing about him. That was what father did sing about. (L.V.15.5100–01)

It seems that two kinds of song are implied in this passage, as there are among the Ju/'hoansi. First, there were the 'traditional' (continually varied) unrecorded rainmaking ones that !Nuin-/kúiten had formerly taught Xa-ttin. Then there was the more personal song that Xa-ttin sang about the deceased !Nuin-/kúiten.

'The Broken String': Versification and Form

As I have pointed out, 'The Broken String' appears in Lloyd's selection of 'Poetry' (Bleek & Lloyd 1911:220–41). 'Songs' may be a more accurate word. She included eleven, but others are to be found in the settings of their original tales; songs were often part of mythical performances. Nine of Lloyd's selected eleven are associated with, or sung by, mythical creatures. Only the remaining two are personal songs. One, '//kabbo's song on the loss of his tobacco pouch', was given by /Han≠kass'o, //Kabbo's son-in-law (Bleek & Lloyd 1911:235). The second, 'The Broken String', was also given by someone other than the original singer, in this case by his son Diä!kwain. Both songs concern personal loss but connect with wider issues.

The exact lineation and punctuation of the song in Dorothea Bleek's published version is shown in the Prologue to this book. It follows vertical pencilled lines still visible in the facsimile phonetic transcription that, probably, Lloyd entered. It follows Lloyd's earlier published version apart from the 'who' at the end of the first line; Dorothea, perhaps more appropriately, placed it at the beginning of the second line. Lloyd's presentation of the song highlights phrases and creates a rhythm that is not in the original (Bleek & Lloyd 1911:237). Here is her lineation:

People were those who
Broke for me the string.
Therefore,
The place became like this for me,

On account of it,
Because the string was that which broke for me.
Therefore,
The place does not feel to me,
As the place used to feel to me,
On account of it,
For,
The place feels as if it stood open before me,
Because the string has broken for me.
Therefore,
The place does not feel pleasant to me,
On account of it.

The manuscript does not observe any regular line or stanza breaks. But the original segments and their varied repetitions can be clearly seen in the phrasing. Major breaks in the text are signalled by the monosyllable *ĩ*. As a verb it means 'to do so, do thus'; it can also mean 'verily, truly, merely, yes' (Bleek 1956:67). It is commonly used to fill in a pause. For instance, in the text leading up to the song, *ĩ* appears seven times, usually between commas, without it being reflected in Lloyd's translation.[9] Once she translated it as 'on account of it' (5096). In the song itself, it appears three times, evidently filling the gap between variations on the theme. The English verse versions raised 'on account of it' to the level of a poignant refrain.

San Songs

In its form of variations on a theme 'The Song of the Broken String' follows San songs in general. They are often based on oral formulae: the singer improvises on them freely, 'creating of each performance a new entity which is, nevertheless, part of a seamless whole' (Biesele 1975a:177). Other songs recorded in the Bleek and Lloyd Archive follow this pattern. In this regard, 'The Song of the Broken String' is comparable to the myths I have discussed: there is no definitive version, and performers can vary the material to suit themselves and the occasion. According to Nicholas England,

> each [Ju/'hoan] song will usually have at least one key phrase that recurs (or better, may occur) intermittently throughout any performance of that piece. Such phrases of course carry more than their basic word mean-

> ings; they are somewhat similar to the '"catch words", crammed full of meanings and surrounded by an extensive halo of implicit meanings' in Mohave shaman songs. (1968:343–44; original parenthesis)

Comparison of Ju/'hoan songs with those of the /Xam suggests that a similar repetitive structure with 'catch words' crammed with meaning was also characteristic in the south. These 'catch words' are nuggets.

England (1968) also found that, although most San music was to some extent vocal, in many instances the music was more important than the words. Writing about 'medicine songs' (which would be akin to !Nuin-/kúiten's rainmaking songs), England (1968:393) makes a highly significant point: he says that they are 'musical manifestations of supernatural power'. Potency is in the sound, as it was in the case of the bag that was 'singing' in the hole (Chapter 6). At a medicine dance, all present are enveloped in the intense, rhythmic sound of the singing and clapping.

Biesele (1975a) outlined a most perceptive approach to this kind of music in her account of a Ju/'hoan man's thumb piano music that she heard in 1971 in western Botswana. The thumb piano itself was at that time considered a fairly recent adoption by the San (Colour Plate 10; England 1968:219). The player in question was /Ai!ae. He was universally known as Jimmy, but following anthropological usage for protecting individuals at the time, Biesele gave his name as '!Kaha, universally known as Jack' (Biesele, pers. comm.). His texts were longer than those of other players. He was

> the only person who uses the thumb piano music as other San use dance music, as a vehicle for trance and as a medium for speaking to God. . . . [He] is different from other men, even from other shamans. . . . [He] does not trance primarily in dancing as do many others. He trances most often while playing the thumb piano. (Biesele 1975a:173)

This is an example of a person trancing outside the formal dance. In that state, the man claimed 'to travel outside his body to check up on the health of relatives who are far away', as San trancers routinely do (Biesele 1975a:183). People accepted that he could enter the spirit realm outside of a trance dance.

Biesele sums up his songs by saying that they have a triple function: they pray to God, relate to other people what God replied, and lament 'his own outcast state among humanity' (Biesele 1975a:174). Here we have an insight that helps us to understand 'The Broken String', a song that relates to the spiritual dimension of San life (rainmaking) and, at the same time, laments personal loss. Although personal songs of this kind are rare among the Ju/'hoansi, England (1968:453) did manage to record one; he called it 'Mourning for the Dead'. He did not hear it subsequently. The repertory of Ju/'hoan songs changes over time as new ones are added and older ones fall away.

Speaking of 'Jack's' songs, Biesele makes points that are fundamental to any understanding of San texts, especially if they are myths or songs, but also some accounts of daily life:

> The words . . . are often extremely elliptical and could be very puzzling for someone unfamiliar with Zhū/twāsi [Ju/'hoan] language and culture. As we get closer to the cultural concepts referred to by the sparse texts, however, it becomes obvious that a rich world of meanings is invoked for a !Kung listener by these allusions. In order for us to experience the songs in anything approaching such meaningfulness, we need a great deal of background. (Biesele 1975a:176)

Biesele's final sentence cannot be overemphasised. It matches points that I made in Chapter 1 and subsequently in discussing the Mantis myths. Complex allusions are often triggered by single words—nuggets—that can easily pass unnoticed by a Westerner who adopts a 'philosophical' approach to San myths. Biesele (1975a:176) emphasises that 'it is extremely hard to understand what is going on in the singing'. We should expect the same level of difficulty in dealing with 'The Song of the Broken String'. An example that she gives of this sort of difficulty is worth quoting in full because it relates to a pivotal word in the song:

> Words in Zhū/twā songs are often altered by phonemic transformation and by syllable elision. Furthermore, the beginning of a phrase may be uttered and broken off, but the entire phrase must nevertheless be inferred from context. An example is Jack's *A tshesī ba ku kwe . . .* ? Which implies: *A tshesī ba ku kwe guni mi?* (What things thus [hound me]?), a phrase commonly heard as a complaint in daily life. *A tshesī* itself is metaphorical. 'What things?' it asks: its referent is not an open question but specifically the *//gaũwasi*, the ghosts or spirits of the dead. It is safer, in general, not to name them. Thus the sentence functions for a member of

the culture as a rhetorical question so obvious that no answer is required. (Biesele 1975a:176–77; all parentheses are Biesele's)

Biesele here invokes the 'taken-for-granted' factor that we have seen often inhibits outsiders' understanding of myths—'so obvious that no answer is required'. Biesele's insight pertains to words in 'The Broken String' that need especial explication: 'people' and 'string'.

'People'

Who, then, were the 'people' who 'broke for me the string'? This must be our interpretative starting point. At least two possibilities may be considered. The first is that 'people' may refer to enemies, in the sense of the Boers who shot !Nuin-/kúiten or other /Xam who were antagonistic. The first of these possibilities is the popular modern reading that refers to the destruction of the /Xam land by the colonial advance.

The second meaning is, however, probably the principal indigenous one, the one that /Xam listeners would have supplied. We have seen that Diä!kwain spoke of how !Nuin-/kúiten went about at night in the form of a lion to protect his own people from 'the other people [*!ke*] who walked about; those who wanted to kill people'. These 'other people' went into people's houses: 'They went to them (to the house) by night, they took a person out of the house. . . . The other people were not used to go nicely about, for, they had continued to go about doing harm. They did not go nicely about, that they m[t] peaceably return home' (L.V.15.5087–89).

It seems clear here that Diä!kwain is speaking about spirits, either of the dead or of malevolent *!gi:ten*, who themselves later returned to their own bodies and homes. If we bear in mind Biesele's comments on '*tshesĩ*', it is likely that Diä!kwain used 'people' (*!kĕ*) because it would be rash to speak out loud about the spirits. In other contexts, Diä!kwain spoke of */nu:-!ke*, spirits of the dead (Bleek 1935a:35–37). In this phrase, the usual word for 'people' is prefaced by the qualifier */nu:*, which signifies 'dead, departed, spirit' (Bleek 1956:350). In such a context, *!ke* can still be translated as 'people'. This, I argue, is a parallel to 'Jack's' use of *tshesĩ* to avoid saying *//gaũwasi* (spirits).

This sort of avoidance was common practice. It is also found in a rain-making song that /Han≠kasso gave Lloyd. Instead of referring directly to 'the dead men [*/nu/nuken !ke*] who are with the rain' (spirits), the singer of the song used the avoidance phrase 'O gallopers', an onomatopoeic reference to thunder (Bleek 1933:304). *Kauüü* in the song is a form of *káokao* or *káukau*, 'to run or gallop' (Bleek 1956:80), with the final syllable repeated to convey the rhythm.

In the light of this evidence, I argue that it was antagonistic *!gi:ten* (dead or living shamans) who interfered with !Nuin-/kúiten's rainmaking practices. This is the burden of the song *in its indigenous context*, not his death at the hands of the Boers. Indeed, he seems to have made rain after his death—up until the time when spirits broke his 'string'.

The song thus invokes a common theme in San thought to which I have repeatedly referred: conflict within the supernatural realm. We encounter it again and again. Diä!kwain, for instance, spoke of such conflict when he said that a *!gi:xa* named !Kwarra-an had said that 'the sorcerers [*!gi:ten*] are used to say that they will take away my power [*!gi:*]' (L.V.4.4179). Similarly, in 'The Song of the Broken String', 'sorcerers' took away !Nuin-/kúiten's power by breaking the 'string'. Consequently, the place became 'unpleasant'.

This reading of 'people' as spirit-world *!gi:ten* makes sense when we proceed to the notion of breaking the string, clearly a supernatural event, although there may also be some conflation of antagonistic spirits and real people. What was this 'string'? There are a number of possibilities.

The 'String'

Was the 'string', as indeed the /Xam word suggests, a bowstring (Brown 1998: 69–70)? Did the breaking of !Nuin-/kúiten's rainmaking bowstring ruin his ability to make rain? I argue that this is the most probable meaning. Or was it one of the 'threads of light' that Ju/'hoan San *!gi:ten* routinely report seeing when they are in trance and that are repeatedly depicted in San rock art (Chapter 7)?[10] Or was it an abstract, emotional bond between him and Diä!kwain's father? Jeremy Hollmann suggests that 'the string is part of the network that connected *!gi:ten* to each other, as well as to rain-animals and game, in extraordinary ways' (2004:276).

The song says: 'People broke *for me* the string'. Does this mean 'people' broke *my* string? Or 'people' broke !Nuin-/kúiten's string and *I* (the singer) suffered as a result? Either way, the sense of personal loss emerges in the pronouns I have italicised.

Note that the text of the song says that 'people' broke the string and caused the place to be 'open'. There are thus two relevant incidents in the overall narrative. One is that the Boers killed !Nuin-/kúiten, an event not mentioned in the song itself. The other is the breaking of the 'string'—the burden of the song. The song is therefore most likely about the end of Xa-ttin's inherited rainmaking skills. Formerly, he had maintained a rainmaking musical bow 'string' that kept the veld 'pleasant'. Now that !Nuin-/kúiten is dead he can no longer protect the 'string' from marauding spirits.

A key passage here is the following one (Diä!kwain is speaking; the parentheses and emphasis are mine):

> [5100] At another time [in the future] father [Xa-ttin] must sing the songs which he [!Nuin-/kúiten] had taught him [over the previous years], father must [5101] sing *about him* [in a song that father would compose?]. That was what father did sing about, he said [that is, sang]: [the song itself follows].

Here we see that Diä!kwain's father would sing the rainmaking songs that !Nuin-/kúiten had taught him. In addition, Diä!kwain's father would sing '*about* him' (that is, about !Nuin-/kúiten). Presumably, this would be a song that Diä!kwain's father himself would compose about !Nuin-/kúiten. The next sentence ('That was what father did sing about, he said') suggests that 'The Song of the Broken String' was the father's own composition. The 'people' who broke the string must have been spirits. The breaking happened in the spirit world: that reading holds good if it was a rainmaking thong for catching a rain-animal *or* a bowstring for calling up the rain. If the 'string' was a 'thread of light', then, too, it was probably malevolent spirits who broke it *in the spirit realm*; it is there that such 'threads' are encountered. But, all in all, it seems to me that Lloyd's alterations in the manuscript that changed 'thong' to 'string' suggest that a musical bowstring was intended, not a thong for catching rain-animals.

What was the effect of the breaking string? Some readers suggest that the 'place' being now 'open' refers to the loss of contact with !Nuin-/kúiten—loneliness. This reading is understandable, but Lloyd originally used the English word 'empty'. Bearing in mind the rainmaking affinities of the song, and that the breaking of the 'string' by malevolent spirits probably heralded the ruin of a rainmaking bow, I suggest that, in this rainmaking context, 'stood open [*bbōken*]' means 'barren, parched', as a result of drought.

Undoubtedly the text of the song is complex—far more complex than an initial reading of an English translation may suggest. Like so many San texts, it is full of ambiguities and pregnant phrases. Bearing in mind the points I have made, I situate the song in its textual context and suggest the following sequence of events:

> !Nuin-/kúiten was a rainmaking shaman who also went about at night in the form of a lion so that he could protect his people from marauding spirits. He realised that he was quick tempered and was afraid that he might kill one of his own people if they troubled him. While on such an expedition in the form of a lion, he killed a Boer farmer's ox. Being a lion, he could not help himself (L.V.15.5084–86). The Boer, who was unaware of !Nuin-/kúiten's lion transformation and who saw and thought of !Nuin-/kúiten only as a human being, mounted a commando, pursued him by following his human tracks, and mortally wounded him. He staggered back to Diä!kwain's father, Xa-ttin. He told Xa-ttin that he wished to teach him his rainmaking songs before he died. He even desired to take Xa-ttin with him to the spirit realm and there teach him the rain songs—but he realised that he could not do this.
>
> After !Nuin-/kúiten had died, he continued to make rain as a spirit. And Xa-ttin sang the rain songs that !Nuin-/kúiten had taught him. Then malevolent spirit 'people' broke !Nuin-/kúiten's rainmaking bowstring. As a result, Xa-ttin could no longer call on him to make rain, and the veld became parched.
>
> Later, Xa-ttin composed a song about the loss of !Nuin-/kúiten. He may have inferred from a prolonged drought what had happened to !Nuin-/kúiten's 'string' in the spirit realm. The Boers may thus have been *ultimately* responsible for the whole catastrophe.

Synthesis

However we read it, close analysis of 'The Song of the Broken String' shows that the song deals with a theme that recurs in /Xam, indeed San, mythology and thought: supernatural conflict and how it can affect daily life. The conflict could be between human beings and spirits or between spirits (or groups of spirits) in the spirit realm. It should come as no surprise that this sort of conflict is also encountered in San rock art (Chapter 7). One conclusion that we can draw is that many narratives and the painted images both engaged with central issues, but each in its own way. The link between myth and image lies in nuggets: both media draw on a common fund of concepts. They are linked by a network of interrelated beliefs and practices.

The ambiguities that we encounter in 'The Song of the Broken String' probably derive in some measure from the way in which the text was recorded and translated. It seems likely that Diä!kwain himself was clear on many of the points that puzzle us today. Or, perhaps more likely, some of the ambiguities that we discern were not, for the San, ambiguities at all. Like the ambivalence of the Mantis in his various escapades, ambiguity was simply a 'fact of life'. The San accepted ambiguity in myth and in daily life without question. Is that a man or a lion? Was the Mantis good or evil? Questions like these would not have occurred to the /Xam, certainly not in the way that much Western philosophy would pose them. The major structural opposites of pure 'good' and 'evil' as a foundation for belief and action were not part of San thinking.

CHAPTER TEN

'They Do Not Possess My Stories'

My place is the Bitterpits.
(//Kabbo, B.II.358)

To situate my approach to myth in a social, everyday context I begin by asking a question that lies behind all such studies: why are some narratives ephemeral while others are preserved over generations and eventually come to be considered as 'myths'? This question leads us toward a central function of many, although not necessarily all, myths throughout the world.

Perhaps baffled by the mystery of myth, Claude Lévi-Strauss (1972) countered with a two-part explanation that sounds a bit like a myth itself, at least in its second part (Geertz 1988). First, he proposed that myths exhibit rather than resolve contradictions inherent in society. That in itself is not a surprising idea. Inevitably, tales reflect, or sometimes invert (and so draw attention to), the social conditions in which they exist. We should, however, be more exact and ask precisely whose social conditions. Those conditions, especially inequality, vary through a community. Hero myths, such as those discussed by Joseph Campbell (1949), are elitist and, in a broad sense, colonialist: the narratives invite hearers to associate themselves with the hero and to fill in the identities of those for whom the abhorrent vanquished stand. Myths that present, as hero myths do, just one dominant point of view triumphantly paper over social inequalities and discord. The Mantis's conflicts with his affines are but one example. They were part of the inherent unpredictability and ambiguity of San social and economic life. But we tend to get only the Mantis's family's view of the matter, not what the affines (Lions, Meerkats) thought about the conflict overall. How did they view the Mantis? This central positioning of the Mantis, in effect a shaman, was clearly significant. At

---

Myth and Meaning: San-Bushman Folklore in Global Context by J. D. Lewis-Williams, 201–209

the same time I allow that the theme of conflict that I have traced through my chosen myths is not necessarily the only possible reading of them. They are multilayered, and their 'meaning' was in all probability contingent on the occasion and manner of performance.

Second, and more intriguingly, Lévi-Strauss believed that myths somehow think themselves through the minds of people: 'I therefore claim to show not how men think in myths, but how myths operate in men's minds without their being aware of the fact' (Lévi-Strauss 1969:12). In this celebrated statement he appears to suggest that myths have an almost mystical autonomy. We are reminded of //Kabbo's words that I placed at the beginning of Chapter 1: 'A story is the wind. It does float along to another place' (L.II.32.2887). So, at least, it seemed to him. In literate societies (and some others as well), the idea of autonomy is taken further. Canonical myths are believed to come from God, no matter who wrote them down. Myths thus seem to be elevated above groups of people and their sectional interests. The origins of the narratives are disguised, and they come to reside in an unquestionable realm in which they simply 'float along': no one (it is believed) can be held responsible for them. From that vantage point, myths may well affect people's minds without their being fully aware of all the myths' implications; myths thus seem to be divine in origin.

Some of the nuggets that I have tried to unpack probably stirred some San minds and emotions without their significances being fully articulated every time people heard them. But that is something different from what Lévi-Strauss seems to be saying. The nuggets that I have identified are the 'hidden' charges that work in people's minds. They are the 'elements' with specifically San associations of which Megan Biesele wrote: although 'merely mentioned in passing', they may be social clues 'with enormous ramifications' (Biesele 1996:145). Comparable in some ways to words with rich connotations in Western poems, these clues, I suggest, intensified the ambience and significance of a myth by invoking otherwise unarticulated beliefs and experiences in an affective penumbra. In this way, oblique phrases and words struck deeper than the apparently prosaic narratives in which they were embedded. Having now examined a number of San myths, we can go further than this observation. As the succession of nuggets in a narration connected one with another and added up, the whole became greater than the sum of the parts. The manifest significance of the basic story became encased in an aura of deeply important, interrelated associations. As Robert Schumann said of Chopin's mazurkas and polonaises, nuggets 'are like cannons hidden beneath flowers'.

All this about nuggets may be true, but we are left with the question why myths are told and retold by *actual people*. Myths do not select, shape, and articulate themselves, as Lévi-Strauss implied. It is individuals and groups of individuals who choose which myths to perform and, through repeated performance, to preserve. They also choose the social circumstances in which they recount myths and thus guide the effect of their performances on their auditors. We therefore need to keep asking a more practical question, one that I have already raised: who benefits from the retelling and reproduction of a myth? Do *all* people in a society benefit equally from the recounting of myths, as the narrators may imply? In any event, what constitutes 'benefit'? In the case of the San the answers to these questions mesh with the significances of many of the nuggets in the myths I have selected. All in all, if we can see past the contradictions in plot between the differing variants of, say, 'The Mantis makes an eland' and 'The fight with the Meerkats', we shall be able to uncover beliefs and rituals that routinely animated contacts between the San's daily lives and an ever-present supernatural realm. I argue that these beliefs underwrote not just San religion but also the social fabric of San society as it existed at any given point in history. It is in beliefs about those contacts, many conflictual, and the ways in which they are effected that we find the reason for the existence and retelling of myths that feature the adventures of the Mantis. Someone benefitted from those Mantis myths, or at any rate benefitted in a more immediate way than many others did, even if 'only' in prestige and influence.

Simultaneously, people sometimes come to accept myths that underwrite their own subservience. Early on, Bronislaw Malinowski (1926) argued that some myths function as charters that validate the *status quo* in societies. A dominant group frequently constructs, places, and reproduces myths of social discrimination so that many of the 'underdogs' also accept the position. Not all, of course, because myth can be a site of contestation, as the 'myths' of history and social memory are in the contemporary West. Nevertheless, Malinowski had a point, at least for some myths. I suspect that the power of myths to persuade people of their authenticity, even when to their disadvantage, lies partly in nuggets that not only support but also substantiate and seemingly validate narratives at deep levels. By subtly invoking networks of meaning they proclaim their own authenticity and that of the whole (apparently immutable) thought-world of the society in which they exist. It is only rare individuals who can step outside all those taken-for-granted parameters.

The approach to myth that I have adopted in this book has, I believe, implications that extend beyond //Kabbo, the San, and the Bleek family. The ways in which San myths work through nuggets help us to develop a more sensitive approach to other cultures around the world and, at least in some measure, to escape the trammels of contemporary Western social and psychological theories that too often govern rather than facilitate our approach to myth in general. The San's complex and, to my way of thinking, often touching narratives, studded as they are with nuggets, open up vistas on lives that were lived long ago and beliefs that, far from being the naive misconceptions of 'primitive' people, were as complex and subtle as any we may encounter elsewhere in the world.

What, then, of the social context in which nineteenth-century San Mantis myths were told?

The long-held belief that the San, in general but allowing for diversity (Guenther 1999:32), are overwhelmingly egalitarian is, I believe, only partially correct (Lewis-Williams & Pearce 2004a, 2004b). This perhaps controversial assessment, one that I have developed in preceding chapters, may seem to run contrary to the usually held view of the San as an exemplary egalitarian, classless, hunter-gatherer society. In general theoretical discourse, numerous writers have sought the origin of social classes and class conflict in such so-called classless societies —where the roots of inequality must surely lie (for example, Bender 1989; McGuire 1992). By and large, there are two not necessarily mutually exclusive explanations. One examines kinship 'interest groups' and the concept of surplus labour as a basis for inequality that is not posited on brute strength (for instance, Gilman 1984). It sees unequal control over resources and labour as an origin of classes. Another explanation posits unequal access to ritual knowledge as a foundation for social inequality.[1] It is less material than the first: the economic base, it contends, does not entirely determine the ideological superstructure. The second of these two explanations fits the San.

Ideologically (in the Marxist sense of masking beliefs) San society is indeed egalitarian. That is why Susan Kent wrote of the San 'façade of egalitarianism' (1993:480). The San are egalitarian in ethos but not necessarily in practice. As early Western travellers and also later anthropologists found, the San in unacculturated circumstances did not have chiefs.[2] In the 1950s, leadership among the Ju/'hoansi was situational (Biesele 1978). In some pivotal situations, effective shamans were more respected (not necessarily more wealthy) than others.[3] On the other hand, in a hunting context an experienced and successful hunter could come into his own. Among the highly acculturated Nharo of the Kalahari, Guenther found: 'The amount of social importance and prestige assigned to trance dancers varies from dancer to dancer. . . . there are highly renowned dancers, whose prestige and influence approximates that of a western superstar athlete or film actor' (Guenther 1986:262, 263). These famous dancers 'command high fees paid in cash, clothing, goats or donkeys' (Guenther 1986:263). They are, however, required to remain modest. Modesty, too, masks inequality.

In more isolated hunter-gatherer communities than the Nharo, the transmission of ritual knowledge may, as Stephen Shennan argues, be 'the only legitimate locus for the generation of inequality. . . . As a result of his mediating position with the spirit world, the shaman has power, privileges and position which are unavailable to anyone else' (2002:223, 224). That is why !Nuin-/kúiten was anxious to pass his knowledge and rainmaking songs on to, specifically, Xa-ttin: they were not open to everyone (Chapter 9). Indeed, the special status of /Xam shamans is evident in numerous passages in the Bleek and Lloyd Archive. /Xam shamans, living as well as dead, had the potential to be feared—which is in itself an indication of inequality (Bleek 1935:14, 22, 32, 37). Although ordinary people had direct access to the Mantis and could ask him to 'bring a male gnu under my darts' (Arbousset & Daumas 1846:256), San shamans were able to convince people that their 'symbolic labour' could create conditions in which ordinary people's daily labour could be successful (Lewis-Williams 1982). It was they who controlled the weather, guided antelope to ambushes, removed economically debilitating sickness, and so forth. They, too, by out-of-body travel dramatised the intercamp social relations and kinship ties that constituted the economic foundation of their communities—in Marxist terms, 'the social relations of production'. Certain myths were an ideological component that contributed to the reproduction of those relations, beliefs, and activities.

In performing this reproductive task, myths and the sort of rock art images that I discussed in Chapter 7 do not merely 'reflect' society, neither directly nor inversely: rather, they act on society. Numerous, certainly not all (Lewis-Williams 1998), San myths and images reinforced the social position of shamans by presenting, even in an amusing, ludic context, supernatural experience and dreaming as a bulwark against malign supernatural influences. Here is evidence for the social classificatory function of myth of which Emile Durkheim and Marcel Mauss long ago wrote: narratives and rock art images deal with 'us', 'them', 'animals', and supernatural beings—all socially constructed, multicomponent categories, not monolithic givens. More specifically, certain myths naturalise and legitimise San kinship categories, their internal contradictions and their relation to scarce resources. Beyond that, the Mantis myths focus specifically on the actions of shamans.

So who benefited from the type of conflict resolution that the San myths and images I have discussed posit? The answer is twofold. In the first place, it was shamans, who had access to the spirit realm and to transformations that spiritual access implied. Their prestige was thereby enhanced. The Mantis, we should not forget, was the *ur*-shaman. But, significantly, the community as a whole could participate in both tales and painted panels. *Everyone* was said to be 'of' the potent eland, not just the shamans (Chapter 8). In Biesele's phrase, everyone present at a trance dance experiences 'an uplifting energy which they feel to be a necessary part of their lives'. The affective impact of the dance on all present, the upliftment, contributes to the prestige of those who are bridges between realms and purveyors of protection.

All in all, the myths I have chosen underwrote the influence of shamans in San society. But that is not the whole story. Any given performance of one of those tales was probably differently received by different listeners. A shaman would perceive a tale differently than would someone who was not a shaman. Some auditors may, in their daily lives, have resented the respect paid to some or even all shamans, even if that respect was restricted to certain spheres of influence and sometimes challenged. We have seen that shamans who failed to make rain were not immune to challenge. There was give and take. People are not insensate pawns moved across the chequerboard of life by binary oppositions, mythemes, or nuggets. Some individuals are aware that they can manipulate the rules, resources, and cognitive frameworks of their communities to their own advantage (Bourdieu 1977; Giddens 1984). This does not mean that they are cynical atheists, although they may be: they are simply turning the system to their own advantage.

We must conclude that, even in their San context, myths were complexly received. I repeat: there were, of course, other myths that dealt with other issues; they, too, are worthy of study. I do not pretend to have presented the whole of San mythology, only aspects of central parts of it that have, I believe, been largely overlooked. As Guenther has clearly put it, the shamans' trance dance is 'the central ritual of the Bushman religion and its defining institution' (1999:181). To ignore statements like that from researchers who have lived with the San would be perverse.

That said, we may confront a final issue. Researchers have noted that the Bleek family does not seem to have paid much attention to conditions on the frontier (Hewitt 2008:32). Were Bleek and Lloyd indifferent to those conditions? They would have been aware of, for instance, Louis Anthing's official 1863 report to the Governor of the Cape Colony on the plight of the San and would have known the author well. Anthing found an appalling situation, and his report split public opinion (de Prada-Samper 2014). In the face of colonial conservatism, he concluded that the depredations of the white settlers meant that the San 'must either steal or perish' and that parties of colonists 'were in the habit of going out to hunt and shoot any Bushmen they might find' (Anthing 1863:7, 11). In the year of his report, Anthing introduced Wilhelm Bleek to three San men whom he brought to Cape Town to stand trial. Presumably in the light of Anthing's testimony, the men were released (Skotnes 2007:64). Earlier, the Swedish naturalist and explorer Anders Sparrman had written: 'Does a colonist at any time get in view of a Boshies-man, he takes fire immediately, and spirits up his horse and dogs, in order to hunt him with more keenness and fury than he would a wolf or any other wild beast' (1789:2:104).

There can be no doubt that the Bleek family was aware of these conditions. Nor were they indifferent to them. In the face of public opinion, they obtained and published sympathetic accounts of their /Xam teachers' lives and hardships. !Nuin-kúiten's death at the hands of a colonial farmer is but one (Chapter 9). //Kabbo summed up his life and indeed those of all San when he said that his home was the Bitterpits—a waterhole filled with brackish, salty water.

There may be another factor at work here. Linking the twentieth-century Kalahari San and the nineteenth-century southern San, Guenther (1989:152) found that the 'Nharo's and /Xam's aesthetic and cosmological preference for fiction rather than fact' may have led the /Xam narrators themselves either to neglect real-life events or to 'fictionalize them in the garb of the old animal tales'. Animals took on new meanings. The Lions and the Meerkats may, in some contexts of narration, have been conflated with white settlers and other farming people. How the /Xam's engagement with encroaching Korana and European settlers was reflected in their folklore is currently being researched (for example, McGranaghan 2014a, 2014b; Wittenberg 2014).

In the tragic circumstances of the 1870s, Lloyd found that //Kabbo 'much enjoyed the thought that the Bushman stories would become known by means of books' (Bleek & Lloyd 1911:x). Why did this please him? In telling his tales to the Bleek family and knowing that other Westerners would thereby come to know them, //Kabbo was, I believe, consciously situating San myth in a new arena. Refashioned on the pages of a book, the /Xam tales would challenge the Western hegemony that the Bleek family, despite their humane stance, represented and the foreign power and economy that were, even as //Kabbo was speaking to Bleek and Lloyd, destroying his culture and way of life. His beliefs, he insisted, were worthy of preservation and attention.

Far from being passive informants sitting impotently at the feet of their masters, //Kabbo and other /Xam people were, by narrating their myths and stories of their lives, having an impact on colonial attitudes to indigenous people—even if that impact turned out to be more limited than they would have wished. They were using the Bleek family for their own ends. For them, the telling of myths was an active engagement with their oppressors, the only form of engagement open to them. Their lives and stories would now 'become known by means of books', live on, and eventually challenge and change demeaning colonial stereotypes of them. To some extent this has proved to be the case, although again not as powerfully as they and we would have wished. Still, as I pointed out at the beginning of Chapter 1, the new South African national motto of reconciliation is in //Kabbo's language. That is a remarkable triumph that //Kabbo and the others achieved. In the centre of the coat of arms itself is a San rock art image; the designers duplicated it to show two figures greeting each other (Colour Plate 12). Every day South Africans see, if not understand, the /Xam language as it is inscribed on every coin and emblazoned on government buildings and documents: *!Ke e: /xarra //ke* — 'People who are different come together'.

I conclude by recalling some words that Lucy Lloyd recorded on July 23, 1873. While he was in Cape Town, //Kabbo pined for the Bitterpits and felt estranged from the Bleek household: 'They do not talk my language'. As we have seen, story-telling and visiting between camps were integral parts, indeed the life-blood, of /Xam social and emotional life. But the Bleek family 'visit their like', and //Kabbo was left behind, lonely (Bleek & Lloyd 1911:301, 302). The poignancy of his isolation and what his stories meant to him is captured in a long account that he gave over some days. Using the /Xam generic word *kukummi* to denote what he was saying, he explained that 'when one has travelled along a road, and goes and sits down, one waits for a story to travel to one, following one along the road'. In Cape Town he did not hear his own people's stories. At one point, an alteration in the translation shows that Lloyd decided to give /*ki* as 'hear' instead of 'get', but, as we have seen in various examples, the /Xam word means something more like 'possess'. For the /Xam, *kukummi* were real, valuable, and to be cherished. I give //Kabbo the last word:

> My fellow men are those who are listening to stories from afar, which float along. . . . I am here; I do not obtain stories. . . . [T]he people of another place are here; they do not possess my stories. . . . [A story is] like the wind, it comes from a far-off quarter, and we feel it. . . . I do merely listen, watching for a story, which I want to hear; while I sit waiting for it; that it may float into my ear. (Bleek & Lloyd 1911:301, 302)

NOTES

CHAPTER ONE

1. 'Bushman' is often said to be a pejorative and sexist word, although some San do choose to use it (Guenther 1986b). 'San', today the generally preferred term, is a Nama (Khoekhoe) word (Haacke & Eiseb 2002.) In using it or even 'Bushman', I explicitly reject any pejorative connotations.

2. The slash, or solidus, in the word /Xam indicates a dental click, one of five such sounds that distinctively characterise all San languages (see page 11).

3. Barnard 2003, 2004; Smith et al. 2000.

4. For example, Biesele & Hitchcock 2011; Chennells 2014; Gabototwe 2014; Gordon 1992; Hoff 1997; Le Roux & White 2004.

5. The manuscripts are preserved in the Jagger Library in the University of Cape Town. On the Bleek and Lloyd Archive and the Bleek family see Bank 2006; Bennum 2004; Deacon & Dowson 1996; Deacon & Skotnes 2014; Lewis-Williams 1981, 2002a, 2014; Lewis-Williams & Challis 2011; McGranaghan 2012; Skotnes 1996a; Spohr 1962; Thornton 2014; Weintroub 2013, no date.

6. Skotnes 2007, and on the web. The *Bantu Studies* publications are reproduced and annotated in Hollmann 2004. Otherwise unpublished passages are given in Guenther 1989 and Lewis-Williams 2002a.

7. For biographies of the /Xam informants see Bank 2006; Bennum 2004; Deacon 1988; 1996b:11–39; Lewis-Williams 1981:26–28, 2002a:14–29.

8. For example, Barnard 1992; Deacon & Deacon 1999; Mitchell 2002:223–25, 291–92; Yellen & Brooks 1988.

9. Adhikari 2010; Anthing 1863; Blundell 2004; Challis 2012; Penn 2005.

10. B.XXV.2414. The speaker is not indicated on the covers of Bleek's notebooks; here the Roman numeral indicates the notebook.

11. Bleek 1932:333, 335; 1956:363.

12. Adhikari 2010; Anthing 1863; Penn 2005.

13. Bank 2006; Lewis-Williams 2014; Weintroube 2014.

14. Bank 2006; Bleek 1928a, 1928b; Deacon 1996a:7–8; Weintroub 2013, 2014, no date.

15. On theories of myth see, for example, Csapo 2005; Edmunds 1990; Segal 1999; Yorke 2013.

16. For example, Biesele 1993; Eastwood & Eastwood 2006; Guenther 1989; Hewitt 2008:205–11; Hollmann 2013; Schmidt 1989, 1996.

17. For instance, Bank 2006; Barnard 1992; Biesele 1996; Guenther 1989, 1999; Hewitt 2008; Lewis-Williams 1981, 1992a, 2002a; McGranaghan 2012; Wessels 2010.

CHAPTER TWO

1. Lewis-Williams 1980, 2003; McGranaghan, Challis, & Lewis-Williams 2013; Mitchell & Challis 2008; Orpen 1874.

2. On Stow and his connection with the Bleek family see Dowson, Tobias, & Lewis-Williams 1994; Schoeman 1997; Skotnes 2008; Young 1908.

3. Lewis-Williams 1980, 1981, 2003; Lewis-Williams & Challis 2011; Lewis-Williams & Pearce 2004a; McGranaghan, Challis, & Lewis-Williams 2013; Mitchell & Challis 2008.

4. On the dating of San rock paintings see Bonneau et al. 2011; Mazel 2009a, 2009b.

5. For example, Blundell 2004; Campbell 1987; Challis 2008, 2012, 2014; Jolly 1996; Loubser & Laurens 1994; Mallen 2008; Mazel 1992; Mitchell 2009; Pearce 2012; Smith 2010; Vinnicombe 1976.

6. For instance, Wilmsen 1989; but see Barnard 2007; Kent 1992; Lee & Guenther 1991, 1993; Mitchell 2002:223–26; Sadr 1997; Solway & Lee 1990.

7. On San-Nguni interaction see Hammond-Tooke 1997, 1998, 1999, 2002; Prins & Lewis 1992.

8. Blundell 2004; Challis 2012, 2014; Mallen 2008.

9. Blundell 2004; Challis 2012, 2014; Pearce 2012:142.

10. Bleek 1874; McGranaghan, Challis, & Lewis-Williams 2013; Orpen 1874.

11. Lewis-Williams 1981, 1992a, 2015; Lewis-Williams & Biesele 1978.

12. Lewis-Williams 1972 conference presentation, published Lewis-Williams 1975:414.

13. Lewis-Williams 1977, published 1981; see also Lewis-Williams & Biesele 1978.

14. Barnard 1992; Biesele 1993, 1996; Guenther 1999; Schmidt 1996.

15. For a discussion of Isaac Schapera's pioneering work, see Guenther 1989:30–36.

16. Biesele 1978; Lee 1979:343–50; Marshall 1976:156–242.

17. *Hxaro* in the Ju/'hoan language; Barnard 1992; Marshall 1961, 1976; Wiessner 1977, 1982.

18. Barrow 1801:1:274; Biesele 1978; Bleek 1924:viii; Blundell 2004; Dowson 1994; Lee 1979:343–50; Leslie 1828:79; Marshall 1976:156–200; Silberbauer 1965:73; Stanford 1910:435; Stow 1905:33; Tanaka 1969:16.

19. 'Khoisan' comprises the hunter-gatherer San and the cattle-herding Khoekhoe; both speak a number of distinctive click languages.

20. For example, Guenther 1999:180–98; Katz 1976, 1982; Lee 1967, 1968, 1979, 1993; Low 2008:86–109, 167–227; Marshall 1969, 1999; Silberbauer 1981:175–78; Valiente-Noailles 1993:182–84.

21. On the specifically San form of shamanism and their rock art see Lewis-Williams 1980, 1981, 1992a, 2001b, 2012; Lewis-Williams & Challis 2011; Lewis-Williams & Pearce 2004a.

22. Lewis-Williams 1992a; Lewis-Williams & Challis 2011; Lewis-Williams & Pearce 2012.

23. Barnard 1992:84–85; Guenther 1999:95–125; Hewitt 2008; Schmidt 1973; Wessels 2010:93–117.

24. Bleek 1924:11, 12, 16, 33; 1932:235; Bleek & Lloyd 1911:3; Orpen 1874:8.

25. Lewis-Williams 1997; Lewis-Williams & Challis 2011:184–85; Lewis-Williams & Pearce 2004a.

26. For example, Biesele 1975a; Guenther 1999:7; Hewitt 2008; Lewis-Williams 1992a.

27. L.VIII.32.8809; L.VIII.2.6178; Bleek 1956:557). Lloyd translated //Kanndoro as 'tinder box owner' (Bleek 1924:27; Bleek & Lloyd 1911:13).

28. Bleek 1956:28; cf. Marshall 1962:232–33.

29. Bleek 1935a:12, 19, 34; Bleek & Lloyd 1911:113; Orpen 1874; for example, Lewis-Williams & Pearce 2004a: 90, 97–100, figures 5.3, 5.4, 6.11, 7.5, 7.6, 8.5a, 8.5b, 10.1.

30. Borcherds 1861:115; Chapman 1868 (1):65; Ellenberger 1953:98; Kolben 1731 (1):76; Paterson 1789:32–32.

31. Challis 2008, 2012; Hollmann 2004:277–78.

32. Bleek 1935a:14–15; 1956:569; L.V.25.6008.

33. Lewis-Williams 1981:77; Lewis-Williams & Pearce 2012:700.

34. On 'threads of light' see Keeney 2003; Lewis-Williams & Challis 2011; Lewis-Williams & Pearce 2004a, 2009; Lewis-Williams & Dowson 2000.

CHAPTER THREE

1. Csapo 2005:181–261; Lévi-Strauss 1972; Propp 1958.

2. Hewitt 2008; Schmidt 1989.

3. Bleek 1933:387; Bleek & Lloyd 1911:323–24.

4. Guenther 1989:29; Hewitt 2008:194, 200.

5. Guenther 1999:74–75. On the changeover from the Primal Time to present-day conditions see Biesele 1993; Guenther 1999:66ff; Wessels 2010.

6. For a more comprehensive view of the Mantis's family see Guenther (1999:68).

7. Arbousset & Daumas 1846:44; Cumming 1850:1:253; Mackenzie 1883:42; Selous 1893:120; Steyn 1971:297; Vinnicombe 2009:172.

8. For example, Dowson 1992; Fock & Fock 1989; Lewis-Williams 1972, 1974, 1981; Maggs 1967; Pager 1971; Smits 1971; Vinnicombe 1976.

9. For instance, Battiss et al. 1958:61; Eastwood & Eastwood 2006; Lewis-Williams 1972, 1981, 1988; Pager 1975; Parkington 1996, 2003; Vinnicombe 1976; Werner 1908; Wright & Mazel 2007.

10. McGranaghan 2014a:13–15; 2014b.

11. Lewis-Williams 1996, 1997; Lewis-Williams & Pearce 2004a: 112–15.

12. L.II.3.454; Arbousset & Daumas 1846:45; Biesele pers. comm.; Sparrman 1789:2:153, 156; Steyn 1971:291; for more on shoes see Wessels 2010:86–90.

13. Marshhall 1962:239; cf., the G/wi Silberbauer 1965:96.

14. Marshall 1962:239; cf. Bleek 1928:26, 46; Silberbauer 1965:126.

15. Bleek & Lloyd 1911:203; L.VIII.17.7516.

16. Lewis-Williams 1996:124–26; 1997:198–200; Lewis-Williams & Pearce 2004a:51–53.

17. For example, Eliade 1964; Vitebsky 1995:50–51; Whitley 2000:24.

18. Bleek 1874:12; 1933; Lewis-Williams 1981:103–16; Lewis-Williams & Pearce 2004b; Orpen 1874:10.

19. England 1968:402, 483, 578; Marshall 1999:196–97.

20. Marshall 1969:370; see also Biesele 1993:109–10.

CHAPTER FOUR

1. Bank 2006; Deacon 1986, 1996:24–29; Lewis-Williams 1981:26–29.

2. For example, Bleek 1929:309; L.VIII.25.8251–68.

3. On /Xam arrows see Deacon 1992.

4. Bleek 1932:233–40; Lewis-Williams 1981:57–61; Lewis-Williams & Biesele 1978.

5. L.V.15.5099; Bleek 1956:24–25; see facsimile page.

6. Marshall 1962:239, 1969:351; Katz 1982: 94; Biesele, pers. comm., notes that the Ju/'hoansi also use the word *n!ai*.

7. The suffix *-i* signifies duration of action: Bleek 1928/1929:168.

8. Katz 1976:286; Marshall 1962:250, 1969:370, 376.

9. //Kabbo gave 'jacket' as *baitje*, as if it were a word in his language; but this is a form of the Dutch/Afrikaans *baadjie*, 'jacket'.

10. The repeated *b* indicates a strong emphasis on the consonant.

11. For example, L.VIII.16.52 Hahn 1881; Bleek 1928; Bleek & Lloyd 1911; Guenther 1999; Marshall 1999; Schapera 1930; Schmidt 1989; Szalay 2002.

12. On Max Müller and solar mythology see Csapo 2005.

13. Barrow 1801:1:283; Bleek 1924:5; Dornan 1925:165; Stow 1905:112.

CHAPTER FIVE

1. Cf. Hammond-Tooke (1977) on Nguni mythology.

2. Lewis-Williams 1981:83–101; Chapter 7.

3. For example, Eliade 1972; Halifax 1980; Hayden 2003:77ff.

CHAPTER SIX

1. For example, Bryden 1893:356; How 1962:11, 39; Mohr 1876; Steedman 1835:(1)139; Stow 1905:85.

2. Casalis 1889:32; Lichtenstein [1812] 1928:(2)51; Sparrman 1789:(2)96; Steedman 1835:(1)139.

3. See also Harris 1838:61; Roberts 1951:279.

4. L.II.3.470; L.II.8.863'; L.III.16.7429; L.III.17.7526'.

5. Biesele 1993:70–72; Lewis-Williams & Pearce 2004a:53–62.

6. Bleek & Lloyd 1911:331; Guenther 2014; Marshall 1999:50–51.

7. Biesele 1993:70, 67–70; Guenther 1986:263; 1999:81; Marshall 1999:74–76.

8. L.II.19.1702; L.II.36.cover rev.; Bleek 1928:13; Leslie 1828:81; Silberbauer 1965:54; 1981:206.

9. Bleek 1935a:5, 7; Heinz & Lee 1978:213; Katz 1982:102–03; Lee 1967:33; Marshall 1962:251; Orpen 1874:10; Silberbauer 1965:99; 1981:175.

10. Lewis-Williams 1996; 1997; Lewis-Williams & Pearce 2004a:112–15.

11. Biesele 1993:147; Lewis-Williams 1996; 1997; Marshall 1976:168–71; see also the 'Anteater's laws': Hewitt 2008:92–95.

CHAPTER SEVEN

1. Garlake 1995:117; Lewis-Williams 1972:51; Lewis-Williams & Dowson 2000:126–27; Pager 1971:321; Smits 1971; Vinnicombe 1976:364; Willcox 1984:202; Walker 1996:79; Woodhouse 1984:107–09.

2. Willcox 1984; for a response see Lewis-Williams 1983.

3. Dowson 1996; Dowson & Lewis-Williams 1993.

4. For example, Challis 2005; Deacon & Foster 2005; Dowson 1992; Eastwood & Eastwood 2006; Hollmann 2002; 2005; Lewis-Williams 1981; Lewis-Williams & Challis 2011; Lewis-

Williams & Pearce 2004a; Mazel 2011; Mguni 2004; Smith & Ouzman 2004; Walker 1996; Wright & Mazel 2007; Yates et al. 1985.

5. For instance, Lewis-Williams 1981, 2002b; Lewis-Williams & Challis 2011; Lewis-Williams & Dowson 2000; Lewis-Williams & Pearce 2004a.

6. Lewis-Williams 1981; Lewis-Williams & Challis 2011; Lewis-Williams & Pearce 2004a.

7. England 1968:432; Marshall 1969.

8. Lewis-Williams 1981:88.

9. For example, Lewis-Williams & Pearce 2004a:figs. 6.4, 6.5, 7.6, 8.7.

10. For instance, Lewis-Williams 1981:81, 95–97, 98–99; Lewis-Williams & Pearce 2004a:90ff.

11. For example, Lewis-Williams 1981; Lewis-Williams & Challis 2011; Lewis-Williams & Pearce 2004a; Vinnicombe 1976.

12. For example, Lewis-Williams & Dowson 2000:50–53.

13. For instance, Lewis-Williams 1981; Lewis-Williams & Pearce 2004a.

14. For example, Pager 1975.

15. Lewis-Williams 1997; Lewis-Williams & Pearce 2004a:126–28.

16. Keeney 2003:102; Marshall 1999:69.

17. Lewis-Williams 1981:103–16; Lewis-Williams & Challis 2011:94–109; Lewis-Williams & Pearce 2004a:137–58.

18. Keeney 2003; Lewis-Williams et al. 2000; Lewis-Williams & Pearce 2009.

19. Dowson 1989; Katz 1982:109; Lewis-Williams 1981: fig. 20.

20. Lewis-Williams & Dowson 1990; Lewis-Williams et al. 2000; Lewis-Williams & Pearce 2004a:147, 180–81, 200.

21. Lewis-Williams 1998; 2001a; Lewis-Williams & Challis 2011; Lewis-Williams & Pearce 2004a; Wright & Mazel 2007.

22. Lewis-Williams & Pearce 2004a:126–28: fig. 6.6; 2012a:fig. 4; Vinnicombe 1976:fig. 107;

23. Lewis-Williams 1986; Lewis-Williams & Dowson 1990:14–15; Lewis-Williams & Pearce 2004a:102.

24. Lewis-Williams 1987: fig. 4; on San bags see Lewis-Williams & Dowson 2000:fig. 53; Lewis-Williams & Pearce 2004a:120, 126.

25. For example, Rudner & Rudner 1970:152–53, pl. 60; Stow & Bleek 1930:pls. 17, 20, 28, 31, 35–38, 48, 61–62, 66; Vinnicombe 1976:24–28, fig. 15; Willcox 1956:captions to pls. 20–22.

26. For instance, Lewis-Williams & Dowson 2000:146–47; Lewis-Williams & Loubser 1986:275–77, fig. 5.6; Lewis-Williams & Pearce 2004a:fig. 10.1.

27. Cf. title of Richard Katz's book *Boiling Energy* (1982).

28. For example, Dowson 1989:figs 9–11; Lewis-Williams & Dowson 2000:figs 17, 30c, 34a.

29. Shortridge 1934:613; Stevenson-Hamilton 1912:127.

30. Lewis-Williams & Challis 2011:fig. 49; Lewis-Williams & Pearce 2004a:fig. 7.6.

31. Lewis-Williams 2003:fig. 56; Lewis-Williams & Pearce 2004a:fig. 7.4.

32. Lewis-Williams 1981:88; Marshall 1969:363–64.

33. On the significance of horses, see Challis 2008, 2012, 2014.

34. See also Lewis-Williams 2003:fig. 42; Lewis-Williams & Dowson 1990:fig. 8.

35. Lewis-Williams & Dowson 1990; Lewis-Williams & Pearce 2004a:147, 180–81, 200.

36. L.II.3.454; Biesele, pers. comm.; Silberbauer 1981:226; Sparrman 1789:2:153, 156; Steyn 1971:291.

37. A transcription of this page was published without comment in *Claim to the Country* (Skotnes 2007:14). The date and place of the note appear to be Sept. 20, 1911 and the country town Kenhardt. Two informants' names appear in this part of the notebook: Nikki Streep and Kaite Boch, and it is not clear who provided the information contained in the last line. It is frustrating that Dorothea made no further reference to the statement.

CHAPTER EIGHT

1. Sparrman 1789:1:92; see also Burchell 1822:1:218

2. For example, Adhikari 2005; Anthing 1863; Blundell 2004; Challis 2008, 2012; Philip 1828: 1:42–48; Sparrman 1789:1:104; Vinnicombe 1976, 2009; Whitelaw 2009; Wright 1971.

3. Mitchell 2010; Orpen 1874:2.

4. For instance, Blundell 2004; Challis 2008, 2012; Stanford 1910; Wright 1971; Wright & Mazel 2007.

5. For example, Challis 2008, 2012, 2014; Philip 1842:184; Vinnicombe 1976; Wright 1971:10, 27.

6. Chapter 7; Lewis-Williams 1981:88; Lewis-Williams & Pearce 2004a:98.

7. *Report* 1883:409; Potgieter 1955:3; Stanford 1910:439.

CHAPTER NINE

1. For example, Adhikari 2010:94–96; Bennum 2004; James 2001; Krog 2004:13; Solomon 2009; van Vuuren 1994; Watson 1991:59–60.

2. L.V.14.5056ff; Bleek 1935a:15–19.

3. Bleek 1932:57, 61; Lewis-Williams & Challis 2011:136–37.

4. Bleek 1932:341; 1933:305, 311.

5. I am grateful to Mark McGranaghan for reminding me of this passage.

6. L.V.4.4180; Bleek 1936:142, 1935a:21; cf. Orpen 1874:4.

7. Bank 2006; Bennun 2004; Deacon 1996a, 1996b; Lewis-Williams 1981, 2002a; McGranaghan 2012.

8. I am grateful to Mark McGranaghan for this insight.

9. Pages 5084, 5087, 5092, 5092, 5093, 5098, 5099.

10. For example, Biesele 1993:70–72; Keeney 2003; Lewis-Williams et al. 2000.

CHAPTER TEN

1. For example, Aldenderfer 1993; Lewis-Williams 1982; Lewis-Williams & Pearce 2005; Shennan 2002; Whitley 1994.

2. Barrow 1801:1:274; Bleek 1824: unnumbered page; Shaw 1820:28; Silberbauer 1965:73; Tanaka 1969:16; Tindall 1856:26; Wilson 1969:60.

3. England 1968:416; Guenther 1986:255, 1999:42–43, 196; Katz 1982:259–60; Lewis-Williams & Pearce 2004b.

REFERENCES

Adhikari, M. 2005. *The Anatomy of a South African Genocide*. Cape Town: University of Cape Town Press.

_____. 2010. *A Anatomy of a South African Genocide: The Extermination of the Cape San Peoples*. Cape Town: University of Cape Town Press.

Aldenderfer, M. 1993. Ritual, hierarchy, and change in foraging societies. *Journal of Anthropological Archaeology* 12:1–40.

Alexander, J. E. 1838. *An Expedition of Discovery into the Interior of Africa, through the hitherto Undescribed Countries of the Great Namaquas, Boschmans, and Hill Damaras*. London: Henry Colburn.

Anthing, L. 1863. Cape Parliamentary Papers: report A39. Cape Town: Government House.

Arbousset, T., & F. Daumas. 1846. *Narrative of an Exploratory Tour to the North-East of the Colony of the Cape of Good Hope*. Cape Town: A. S. Robertson.

Atkinson, J. M. 1992. Shamanisms today. *Annual Review of Anthropology* 21:307–30.

Bahn, P. 2010. *Prehistoric Rock Art: Polemics and Progress*. New York: Cambridge University Press.

Bank, A. 2006. *Bushmen in a Victorian World: The Remarkable Story of the Bleek-Lloyd Collection of Bushman Folklore*. Ca

Barnard, A. 1986. Structure and fluidity in Khoisan religious ideas. Paper presented at the Second Satterthwaite Colloquium on African Religion and Ritual, Satterthwaite, Cumbria.

_____. 1992. *Hunters and Herders of Southern Africa: A Comparative Ethnography of the Khoisan Peoples*. New York: Cambridge University Press.

_____. 2003. Diverse people unite: Two lectures on Khoisan imagery and the state. *Occasional Papers*, 94. Edinburgh: Centre of African Studies, Edinburgh University.

_____. 2004. Coat of arms and the body politic: Khoisan imagery and South African national identity. *Ethnos* 69:5–22.

_____. 2007. *Anthropology and the Bushman*. Oxford: Berg.

_____. 2013. Cognitive and social aspects of language origins, in *New Perspectives on the Origins of Language*, C. Lefebvre, B. Comrie, & H. Cohen (Eds.). Amsterdam: John Benjamins Publishing Company, 53–71.

Barrow, J. 1801. *An Account of Travels into the Interior of Southern Africa*. London: Cadell & Davies.

Battiss, W., G. H. Franz, J. W. Grossert, & H. P. Junod. 1958. *The Art of Africa*. Pietermaritzburg: Shuter & Shooter.

Beidelman, T. O. 1980. The moral imagination of the Kaguru: Some thoughts on tricksters, translation, and comparative analysis. *American Ethnologist* 7:27–42.

_____. 1993. The moral imagination of the Kaguru: Some thoughts on tricksters, translation, and comparative analysis, in *Mythical Trickster Figures*, W. J. Hynes & W. G. Doty (Eds.). Tuscaloosa: University of Alabama Press, 174–92.

Bender, B. 1989. The roots of inequality, in *Domination and Resistance*, D. Miller, M. Rowlands, & C. Tilley (Eds.). London: Unwin & Hyman, 83–93.

Bennum, N. 2004. *The Broken String: The Last Words of an Extinct People*. London: Viking.

Biesele, M. 1975a. Song texts by the Master of Tricks: Kalahari San thumb piano music. *Botswana Notes and records* 7:171–88.

_____. 1975b. Folklore and ritual of !Kung hunter-gatherers. Unpublished Ph.D. thesis. Harvard University.

_____. 1978. Sapience and scarce resources: Communication systems of the !Kung and other foragers. *Social Science Information* 17:921–47.

_____. 1983. Interpretation in rock art and folklore: Communication systems in evolutionary perspective, in *New Approaches to Southern African Rock Art,* J. D. Lewis-Williams (Ed.). Cape Town: South African Archaeological Society, Goodwin Series 4, 54–60.

_____. 1986. 'Anyone with sense would know': Tradition and creativity in !Kung narrative and song', in *Contemporary Studies on Khoisan, in Honour of Oswin Köhler on the Occasion of his 75th Birthday (Part I)*, R. Vossen & K. Keuthmann (Eds.). Hamburg: Helmut Buske, 83–106.

_____. 1993. *Women Like Meat: The Folklore and Foraging Ideology of the Kalahari Ju/'hoan.* Johannesburg: Witwatersrand University Press.

_____. 1996. 'He stealthily lightened at his brother-in-law' (and thunder echoes in Bushman oral tradition a century later), in *Voices from the Past: /Xam Bushmen and the Bleek and Lloyd Collection,* J. Deacon & T. A. Dowson (Eds.). Johannesburg: Witwatersrand University Press, 142–60.

Biesele, M., & R. K. Hitchcock. 2011. *The Ju/'hoan San of Nyae Nyae and Namibian Independence.* New York: Berghahn.

Bleek, D. F. 1924. *The Mantis and His Friends.* Cape Town: Maskew Miller.

_____. 1928a. *The Naron: A Bushman Tribe of the Central Kalahari.* New York: Cambridge University Press.

_____. 1928b. Bushmen of central Angola. *Bantu Studies* 3:105–25.

_____. 1928/1929. Bushman grammar: A grammatical sketch of the language of the */xam-ka-!k'e. Zeitschrift für Eingeborenen-Sprachen* 19:81–98; 20:161–74.

_____. 1929. Bushman folklore. *Africa* 2:302–13.

_____. 1931. Customs and beliefs of the /Xam Bushmen: Part I: Baboons. *Bantu Studies* 5:167–79.

_____. 1932. Customs and beliefs of the /Xam Bushmen: Part II: The lion. Part III: Game animals. Part IV: Omens, wind-making, clouds. *Bantu Studies* 6:47–63, 323–42.

_____. 1933. Customs and beliefs of the /Xam Bushmen: Part V: The rain. Part VI: Rain-making. *Bantu Studies* 7:297–312, 375–92.

_____. 1935a. Beliefs and customs of the /Xam Bushmen: Part VII: Sorcerors (*sic*). *Bantu Studies* 9:1–47.

_____. 1935b. !Kung Bushman mythology. *Zeitschrift für Eingeborenen-Sprachen* 25:261–83.

_____. 1936. Customs and beliefs of the /Xam Bushmen: Part VIII: More about sorcerors (*sic*) and charms. Special speech of animals and the moon used by /Xam Bushmen. *Bantu Studies* 10:131–62, 163–99.

_____. 1956. *A Bushman Dictionary.* New Haven, CT: American Oriental Society.

Bleek, E., & D. F. Bleek. 1909. Notes on the Bushmen, in H. Tongue, *Bushman Paintings*. London: Clarendon Press, 36–44.

Bleek, W. H. I. 1874. Remarks on Orpen's 'A glimpse into the mythology of the Maluti Bushmen. *Cape Monthly Magazine* (n.s.) 9:10–13.

Bleek, W. H. I. 1875. *A Brief Account of Bushman Folklore and Other Texts*. London: Trübner & Brockhaus.

Bleek, W. H. I., & L. C. Lloyd. 1911. *Specimens of Bushman Folklore*. London: George Allen.

Blundell, G. 2004. *Nqabayo's Nomansland: San Rock Art and the Somatic Past*. Uppsala: Uppsala University Press.

Bonneau, A., F. Brock, T. Higham, D. G. Pearce, & A. M. Pollard. 2011. An improved pre-treatment protocol for radiocarbon dating black pigments in San rock art. *Radio Carbon* 53(3):419–28.

Borcherds, P. B. 1861. *An Autobiographical Memoir*. Cape Town: Robertson.

Bourdieu, P. 1977. *Outline of a Theory of Practice*. New York: Cambridge University Press.

Brown, D. 1998. *Voicing the Text: South African Oral Poetry and Performance*. Cape Town: Oxford University Press.

Bryden, H. A. 1893. *Gun and Camera in Southern Africa: A Year of Wanderings in Bechuanaland, the Kahari Desert, and the Lake River Country, Ngamiland*. London: Edward Stanford.

_____. 1899. *Great and Small Game of Africa*. London: Rowland Ward.

Burchell, W. J. 1822. *Travels in the Interior of Southern Africa*. London: Longman, Hurst, Rees.

Callaway, G. 1969. *The Fellowship of the Veld: Sketches of Kafir Life in South Africa*. New York: Negro Press.

Campbell, C. 1986. Images of war: A problem in San rock art research. *World Archaeology* 18:255–68.

_____. 1987. Contact period rock art of the southeastern mountains. Unpublished M.A. thesis, University of the Witwatersrand, Johannesburg.

Campbell, J. 1815. *Travels in South Africa, Undertaken at the Request of the Missionary Society*. London: Black, Parry & Co.

_____. 1949. *The Hero with a Thousand Faces*. New York: Pantheon Books.

Casali, E. 1889. *My Life in Basuto Land*. London: The Religious Tract Society.

Challis, W. 2005. 'The men with rhebok's heads; they tame elands and snakes': Incorporating the rhebok antelope in the understanding of southern African rock art. *South African Archaeological Society Goodwin Series* 9, 11–20.

_____. 2008. The impact of the horse on the Amatola 'Bushmen': New identity in the Maloti-Drakensberg mountains of southern Africa. Unpublished Ph.D. thesis, University of Cambridge.

_____. 2012. Creolisation on the nineteenth-century frontiers of southern Africa: A case study of the AmaTola 'Bushmen' in the Maloti-Drakensberg. *Journal of Southern African Studies* 38:265–80.

_____. 2014. Binding beliefs: The creolisation process in a 'Bushman' raider group in nineteenth-century southern Africa, in *The Courage of //Kabbo: Celebrating the 100th Anniversary of the Publication of Specimen's of Bushman Folklore*, J. Deacon & P. Skotnes (Eds.). Cape Town: University of Cape Town Press, 247–64.

Challis, S., J. Hollmann, & M. McGranaghan. 2013. 'Rain snakes' from the Senqu River: New light on Qing's commentary of San rock art from Sehonghong, Lesotho. *Azania: Archaeological Research in Africa* 48:331–54.

Chapman, J. 1868. *Travels in the Interior of South Africa*. London: Bell & Daly.

Chennells, R. 2004. Some reflections, in *The Courage of //Kabbo: Celebrating the 100th Anniversary of the Publication of Specimen's of Bushman Folklore*, J. Deacon & P. Skotnes (Eds.). Cape Town: University of Cape Town Press, 417–22.

Csapo, E. 2005. *Theories of Myth*. Oxford: Blackwell.

Cumming, R. G. 1850. *Five Years of a Hunter's Life in the far Interior of South Africa*. London: John Murray.

Curlé, L. C. 1913. Notes on Namaqualand Bushmen. *Transactions of the Royal Society of South Africa* 3:113–20.

Deacon, H. J. 1976. *Where Hunters Gathered: A Study of Holocene Later Stone Age People in the Eastern Cape*. Claremont: South African Archaeological Society Monograph Series 1.

Deacon, H. J., & J. Deacon. 1999. *Human Beginnings in South Africa*. Cape Town: David Philip.

Deacon, J. 1986. 'My place is the Bitterpits': The home territory of Bleek and Lloyd's /Xam San informants. *African Studies* 45:135–55.

_____. 1988. The power of a place in understanding southern San rock engravings. *World Archaeology* 20:129–40.

_____. 1992. *Arrows as Agents of Belief among the /Xam Bushmen*. Maragret Shaw Lecture 3. Cape Town: South African Museum.

_____. 1994. Rock engravings and the folklore, in *Contested Images: Diversity in Southern African Rock Art Research,* T.A. Dowson & J. D. Lewis-Williams (Eds.). Johannesburg: Witwatersrand University Press, 237–56.

_____. 1996a. Archaeology of the Flat and Grass Bushmen, in *Voices from the Past: /Xam Bushmen and the Bleek and Lloyd Collection*, J. Deacon & T. A. Dowson (Eds.). Johannesburg: Witwatersrand University Press, 245–70.

_____. 1996b. A tale of two families: Wilhelm Bleek, Lucy Lloyd, and the /Xam San of the Northern Cape, in *Miscast: Negotiating the Presence of the Bushmen*, P. Skotnes (Ed.). Cape Town: University of Cape Town Press, 93–114.

_____. 2001. A /Xam San conundrum: What comes first, the art or the place?, in *Theoretical Perspectives in Rock Art Research*, K. Helskog (Ed.). Oslo: Novus Forlag, 242–50.

Deacon, J., & T. A. Dowson (Eds.) 1996. *Voices from the Past: /Xam Bushmen and the Bleek and Lloyd Collection*. Johannesburg: Witwatersrand University Press.

Deacon, J., & C. Foster. 2005. *My Heart Stands in the Hill*. Cape Town: Struik.

Deacon, J., & P. Skotnes (Eds.) 2014. *The Courage of //kabbo: Celebrating the 100th Anniversary of the Publication of* Specimens of Bushman Folklore. Cape Town: UCT Press.

de Prada-Samper, J. M. 2014. Louis Anthing (1829–1902), in *The Courage of //Kabbo: Celebrating the 100th Anniversary of the Publication of* Specimen's of Bushman Folklore, J. Deacon & P. Skotnes (Eds.). Cape Town: University of Cape Town Press, 61–69.

Dornan, S. S. 1925. *Pygmies and Bushmen of the Kalahari*. London: Seeley, Service, & Co.

Dowson, T. A. 1988. Revelations of religious reality: The individual in San rock art. *World Archaeology* 20:116–28.

_____ 1989. Dots and dashes: Cracking the entoptic code in Bushman rock paintings. *South African Archaeological Society Goodwin Series* 6, 84–94.

Dowson, T. A. 1992. *Rock Engravings of Southern Africa*. Johannesburg: Witwatersrand University Press.

_____. 1994. Reading art, writing history: Rock art and social change in southern Africa. *World Archaeology* 25:332–45.

_____. 1996. Reproduction and consumption: The use of rock art imagery in southern Africa today, in *Miscast: Negotiating the Presence of the Bushmen*, P. Skotnes (Ed.). Cape Town: University of Cape Town Press, 314–21.

_____. 2000. Paintings as politics: Exposing historical processes in hunter-gatherer rock art, in *Hunter-Gatherers in the Modern World: Conflict, Resistance, and Self-determination*, P. P. Schweitzer, M. Biesele, & R. K. Hitchcock (Eds.). New York: Berghahn, 413–26.

Dowson, T. A., & Lewis-Williams, J. D. 1993. Myths, museums, and southern African rock art. *South African Historical Journal* 29:44–60.

Dowson, T. A., P. V. Tobias, & J. D. Lewis-Williams. 1992. 'Blue ostriches' captured. *Nature* 358:185.

_____. 1994. The mystery of the Blue Ostriches: Clues to the origin and authorship of a supposed rock painting. *African Studies* 53:3–38.

Eastwood, E., & C. Eastwood. 2006. *Capturing the Spoor: An Exploration of Southern African Rock Art*. Cape Town: David Philip.

Edmunds, L. (Ed.) 1990. *Approaches to Greek Myth*. Baltimore MD: John Hopkins University Press.

Eliade, M. 1964. *Shamanism: Archaic Techniques of Ecstasy*. Princeton, NJ: Princeton University Press.

_____. 1972. *Shamanism: Archaic Techniques of Ecstasy*. New York: Routledge and Kegan Paul.

Ellenberger, D. F., & J. C. Macgregor. 1912. *History of the Basuto, Ancient and Modern*. London: Caxton.

Ellenberger, V. 1953. *La Fin Tragique des Bushmen*. Paris: Amiot Dumont.

England, N. 1968. Music among the Ju/wa-si of South West Africa and Botswana. Ph.D. thesis, Harvard University, Cambridge, MA.

Fock, G. J., & D. Fock. 1989. *Felsbilder in SüdAfrica, Teil III: Felsbilder im Vaal-Oranje-Becken*. Köln: Böhlau Verlag.

Freilich, M. 1975. Myths, method, and madness. *Current Anthropology* 16:207–26.

Gabototwe, J. S. 2014. The /Xam and San youth of today, in *The Courage of //Kabbo: Celebrating the 100th Anniversary of the Publication of Specimen's of Bushman Folklore*, J. Deacon & P. Skotnes (Eds.). Cape Town: University of Cape Town Press, 423–28.

Garlake, P. 1995. *The Hunter's Vision: The Prehistoric Art of Zimbabwe*. London: British Museum Press.

Geertz, C. 1973. *The Interpretation of Cultures*. New York: Basic Books.

_____. 1988. *Works and Lives: The Anthropologist as Author*. Stanford, CA: Stanford University.

George, A. 1999. *The Epic of Gilgamesh: The Babylonian Epic Poem and Other Texts in Akkadian and Sumerian*. London: Penquin.

Giddens. A. 1984. *The Constitution of Society: Outline of a Theory of Structuration*. Berkeley and Los Angeles: University of California Press.

Gilman, A. 1984. Explaining the Upper Palaeolithic Revolution, in *Marxist Perspectives in Archaeology*, M. Sprigs (Ed.). New York: Cambridge University Press, 115–26.

Goody, J. 1977. *The Domestication of the Savage Mind*. New York: Cambridge University Press.

Gordon, R. J. 1986. End note: A Namibian perspective on Lorna Marshall's ethnography, in *The Past and Future of !Kung Ethnography: Critical Reflections and Symbolic Perspectives, Essays in Honour of Lorna Marshall*, M Biesele, R. Gordon, & R. Lee (Eds.). Hamburg: Helmut Buske Verlag, 359–74.

_____. 1992. *The Bushman Myth: The Making of a Namibian Underclass*. Boulder, CO: Westview Press.

Guenther, M. 1979. Bushman religion and the (non)sense of anthropological theory. *Sociologus* 29:102–32.

_____. 1986a. *The Nharo Bushmen of Botswana: Tradition and Change*. Hamburg: Helmut Buske Verlag.

Guenther, M. 1986b. 'San' or 'Bushmen'?, in *The Past and Future of !Kung Ethnography: Critical Reflections and Symbolic Perspectives, Essays in Honour of Lorna Marshall*, M. Biesele, R. Gordon, & R. Lee (Eds.). Hamburg: Helmut Buske Verlag, 27–51.

_____. 1989. *Bushman Folktales: Oral Traditions of the Nharo of Botswana and the /Xam of the Cape*. Stuttgart: Franz Steiner Verlag.

_____. 1994. The relationship of Bushman art to ritual and folklore, in *Contested Images: Diversity in Southern African Rock Art Research*, T. A. Dowson and J. D. Lewis-Williams (Eds.). Johannesburg: Witwatersrand University Press, 257–74.

_____. 1996. Attempting to contextualise /Xam oral tradition, in *Voices from the Past: /Xam Bushmen and the Bleek and Lloyd Collection*, J. Deacon and T. A. Dowson (Eds.). Johannesburg: Witwatersrand University Press, 79–99.

_____. 1997. Jesus Christ as trickster in the religion of the contemporary Bushmen, in *The Games of God and Men: Essays in Play and Performance*, K. Koepping (Ed.). Hamburg: LitVerlag, 203–30.

_____. 1999. *Tricksters and Trancers: Bushman Religion and Society*. Bloomington: Indiana University Press.

_____. 2006. *N//àe* ('Talking'): The oral and rhetorical base of San culture. *Journal of Folklore Research* 43:241–61.

_____. 2014. Dreams and stories, in *The Courage of //kabbo: Celebrating the 100th Anniversary of the Publication of Specimens of Bushman Folklore*, J. Deacon & P. Skotnes (Eds.). Cape Town: UCT Press, 195–210.

Haacke, W. H. G., & E. Eiseb. 2002. *A Khoekhoegowab Dictionary with an English-Khoekhoegowab Index*. Windhoek: Gamsberg Mcmillan.

Hahn, T. 1881. *Tsuni-//Goam: The Supreme Being of the Khoi-Khoi*. London: Trübner.

Halifax, J. 1980. *Shamanic Voices: A Survey of Visionary Narratives*. Harmondsworth: Penguin.

Hammond-Tooke, W. D. 1977. Lévi-Strauss in a garden of millet: The structural analysis of a Zulu folktale. *Man* 12:76–86.

_____. 1997. Whatever happened to /Kaggen? A note on Cape Nguni/Khoisan borrowing. *South African Archaeological Bulletin* 52:122–24.

_____. 1998. Selective borrowing? The possibility of San shamanistic influence on southern Bantu divination and healing practices. *South African Archaeological Bulletin* 53:9–15.

_____. 1999. Divinatory animals: Further evidence of San–Nguni borrowing? *South African Archaeological Bulletin* 54:128–32.

_____. 2002. The uniqueness of Nguni mediumistic divination in Southern Africa. *Africa* 72: 277–92.

Harris, W. C. 1831. *Narrative of an Expedition into Southern Africa during the Years 1836 and 1837*. Bombay: The American Mission Press.

Haydon, B. 2003. *A Prehistory of Religion: Shamans, Sorcerers, and Saints*. Washington, D. C.: Smithsonian Books.

Heikkinen, T. 2011. *Hai//om and !Xũ Stories from North Namibia*, S. Schmidt (Ed.). Köln: Rüdiger Köppe Verlag.

Heinz, H.-J. 1966. The social organization of the !Kõ Bushmen. Ph.D. thesis, University of South Africa, Pretoria.

Heinz, H.-J., & M. Lee, 1978. *Namkwa: Life Among the Bushmen*. London: Jonathan Cape.

Hewitt, J. 1931. Artefacts from Melkhoutboom. *South African Journal of Science* 28:540–48.

Hewitt, R. 2008. *Structure, Meaning, and Ritual in the Narratives of the Southern San*. Johannesburg: Witwatersrand University Press.

Hoff, A. 1997. The water snake of the Khoekhoen and the /Xam. *South African Archaeological Bulletin* 52:21–37.

Hollmann, J. C. 2002. Natural models, ethology, and San rock-paintings: Pilo-erection and depictions of bristles in southeastern South Africa. *South African Journal of Science* 98:563–67.

Hollmann, J. C. (Ed.) 2004. *Customs and Beliefs of the /Xam Bushmen*. Johannesburg: Witwatersrand University Press.

_____. 2005. 'Swift-people': Therianthropes and bird symbolism in hunter-gatherer rock-paintings, Western and Eastern Cape Provinces, South Africa. *South African Archaeological Society Goodwin Series* 9:21–33.

_____. 2013. Exploring the Gestoptefontein-Driekuil complex (GDC): An ancient women's ceremonial centre in the North West Province, South Africa. *South African Archaeological Bulletin* 68:146–59.

How, M. W. 1962. *The Mountain Bushmen of Basutoland*. Pretoria: van Schaik.

James, A. 2001. *The First Bushman's Path: Stories, Songs, and Testimonies of the /Xam of the Northern Cape*. Pietermaritzburg: University of Natal Press.

Johnson, T., H. Rabinowitz, & P. Sieff. 1959. *Rock-Paintings of the South-West Cape*. Cape Town: Nasionale Boekhandel.

Jolly, P. 1986. A first-generation descendant of the Transkei San. *South African Archaeological Bulletin* 41:6–9.

_____. 1996. Symbiotic interaction between Black farmers and south-eastern San. *Current Anthropology* 37:277–05.

_____. 2006. The San rock painting from 'The upper cave at Mangolong', Lesotho. *South African Archaeological Bulletin* 61:68–75.

Kannemeyer, D. R. 1890. Stone implements of the Bushmen. *Cape Illustrated Magazine* 1:120–30.

Katz, R. 1976. Education for transcendence: !Kia-healing with the Kalahari !Kung, *Kalahari Hunter-Gatherers: Studies of the !Kung San and Their Neighbours*, R. B. Lee & I. DeVore (Eds.). Cambridge, MA: Harvard University Press, 281–301.

_____. 1982. *Boiling Energy: Community Healing among the Kalahari San*. Cambridge, MA: Harvard University Press.

Katz, R., M. Biesele, & V. St. Denis. 1997. *Healing Makes Our Hearts Happy: Spirituality and Cultural Transformation among the Kalahari Ju'/hoansi*. Rochester, NY: Inner Traditions.

Keenan, J. 1981. The concept of the mode of production in hunter-gatherer societies, in *The Anthropology of Precapitalist Societies*, J. S. Khan & J. R. Llobera (Eds.). London: Macmillan, 2–21.

Keeney, B. 2003. *Kalahari Bushmen Healers*. Philadelphia: Ringing Rocks Press.

_____. 2003. *Ropes to God: Experiencing the Bushman Spiritual Universe*. Philadelphia: Ringing Rocks Press.

Kent, S. 1992. The current forager controversy: Real versus ideal views of hunter-gatherers. *Man* 27:45–70.

_____. 1993. 'And justice for all': The development of political centralisation among newly sedentary foragers. *American Anthropologist* 91:703–12.

Kina-ha. 1877. A contribution from a Bushman. *Orange Free State Monthly Magazine* 1:83–85.

Kirk, G. S. 1974. *The Nature of Greek Myths*. Harmondsworth: Penguin.

Kolben, P. 1731. *The Present State of the Cape of Good Hope*. London: W. Innys.

Krog, A. 2004. *The Stars Say 'tsau': /Xam Poetry of Diä!kwain, Kweiten-ta-//ken, /A!kúnta, /Han≠kass'o and //Kabbo*. Cape Town: Kwela Books.

Lee, R. B. 1967. Trance cure of the !Kung Bushmen. *Natural History* 78(40):14–22.

_____. 1968. The sociology of !Kung Bushman trance performances, in *Trance and Possession States*, R. Prince (Ed.). Montreal: R. M. Bucke Memorial Society, 35–54.

Lee, R. B. 1969. Eating Christmas in the Kalahari. *Natural History* 78(10):14–22, 60–61.

_____. 1979. *The !Kung San: Men, Women, and Work in a Foraging Society*. New York: Cambridge University Press.

_____. 1993. *The Dobe Ju/'hoansi*. New York: Harcourt Brace.

Lee, R. B., & M. Guenther. 1991. Oxen or onions? The search for trade (and truth) in the Kalahari. *Current Anthropology* 32:592–601.

_____. 1993. Problems in Kalahari ethnography and the tolerance of error. *History in Africa* 20:185–235.

Le Roux, W., & A. White. 2004. *Voices of the San*. Cape Town Kwela Books.

Leslie, L. 1828. Some remarks on the Bushmen of the Orange River. *Edinburgh New Philosophical Journal*. April, 79–82.

Lévi-Strauss, C. 1969. *The Raw and the Cooked: Introduction to a Science of Mythology*. New York: Harper & Row.

_____. 1972. *Structural Anthropology*. Harmondsworth: Penguin.

_____. 1978. *Structural Anthropology: Vol. Two*. Harmondsworth: Penguin.

Lewis-Williams, J. D. 1972. The syntax and function of the Giant's Castle rock paintings. *South African Archaeological Bulletin* 27:49–65.

_____. 1974. Superpositioning in a sample of rock paintings from the Barkly East district. *South African Archaeological Bulletin* 29:93–103.

_____-. 1975. The Drakensberg rock paintings as an expression of religious thought, in *Les religions de la Préhistoire*, E. Anati (Ed.). Capo di Ponte: Centro Camuno di Studi Preistorice, 413–26.

_____. 1977. Believing and seeing: An interpretation of symbolic meanings in southern San rock paintings. Unpublished Ph.D. thesis. Durban: University of Natal.

_____. 1980. Ethnography and iconography: Aspects of southern San thought and art. *Man* (N.S.) 15:467–82.

Lewis-Williams, J. D. 1981. *Believing and Seeing: Symbolic Meanings in Southern San Rock Paintings*. London: Academic Press.

_____. 1982. The economic and social context of southern San rock art. *Current Anthropology* 23:429–49.

_____. 1983. Introductory essay: Science and rock art, in *New Approaches to Southern African Rock Art, South African Archaeological Society Goodwin Series*, J. D. Lewis-Williams (Ed.), 4:3–13.

_____. 1986. The last testament of the southern San. *South African Archaeological Bulletin* 41:10–11.

_____. 1987. A dream of eland: An unexplored component of San shamanism. *World Archaeology* 19:165–77.

_____. 1988. People of the eland: An archaeolinguistic crux, in *Hunters and Gatherers: Property, Power and Ideology*, T. Ingold, D. Riches, & J. Woodburn (Eds.). London: Berg, 203–11.

_____. 1992a. Ethnographic evidence relating to 'trance' and 'shamans' among northern and southern Bushmen. *South African Archaeological Bulletin* 47:56–60.

_____. 1992b. *Vision, Power and Dance: The Genesis of a Southern African Rock Art Panel.* Fourteenth Kroon Lecture. Amsterdam: Stichting Nederlands Museum voor Anthropologie en Praehistorie.

_____. 1995. Modelling the production and consumption of rock art. *South African Archaeological Bulletin* 50:143–54.

_____. 1996. A visit to the Lion's house: The structure, metaphors, and sociopolitical significance of a nineteenth-century Bushman myth, in *Voices from the Past: /Xam Bushmen and the Bleek and Lloyd Collection*, J. Deacon & T. A. Dowson (Eds.). Johannesburg: Witwatersrand University Press, 122–41.

_____. 1997. The Mantis, the Eland, and the Meerkats: Conflict and mediation in a nineteenth-century San myth, in *Culture and the Commonplace: Anthropological Essays in Honour of David Hammond-Tooke*, P. McAllister (Ed.). Johannesburg: Witwatersrand University Press, 195–216. Reprinted in Lewis-Williams, J. D., 2002c, *A Cosmos in Stone: Interpreting Religion and Society though Rock Art*, 73–94. Walnut Creek, CA: AltaMira Press.

_____-. 1998. Quanto? The issue of 'many meanings' in southern African rock art research. *South African Archaeological Bulletin* 53:86–97.

_____. 2001a. Monolithism and polysemy: Scylla and Charybdis in rock art research, in *Theoretical Perspectives in Rock Art Research*, K. Helskog (Ed.). Oslo: Novus Forlag, 23–39.

_____. 2001b. Brainstorming images: Neuropsychology and rock art research, in *Handbook of Rock Art Research*, D. Whitley (Ed.). Walnut Creek, CA: AltaMira Press, 332–57.

_____. 2002a. *Stories That Float from Afar: Ancestral Folklore of the San of Southern Africa.* Cape Town: David Philip.

_____. 2002b. *A Cosmos in Stone: Interpreting Religion and Society through Rock Art.* Walnut Creek, CA: AltaMira Press.

_____. 2003. *Images of Mystery: Rock Art of the Drakensberg.* Cape Town: Double Storey.

_____. 2010. The imagistic web of San myth and landscape. *Southern African Humanities* 22:1–18.

_____. 2012a. Rock art and shamanism, in *A Companion to Rock Art*, J. McDonald & P. Veth (Eds.). Chichester: Wiley-Blackwell, 17–33.

Lewis-Williams, J. D. 2013a. From illustration to social intervention: Three nineteenth-century /Xam myths and their implications for understanding San rock art. *Cambridge Archaeological Journal* 23:241–62.

_____. 2013b. Qwanciqutshaa, a nineteenth-century southern San mythological being: Who is he and what does his name mean? *South African Archaeological Bulletin* 68:79–85.

_____. 2014. The Bleek and Lloyd Collection and rock art research: 1874–2011, in *The Courage of //kabbo: Celebrating the 100th Anniversary of the Publication of Specimens of Bushman Folklore*, J. Deacon & P. Skotnes (Eds.). Cape Town: University of Cape Town Press, 131–49.

_____. 2015. Texts and evidence: Negotiating San words and images. *South African Archaeological Bulletin* 70.

Lewis-Williams, J. D., & M. Biesele. 1978. Eland hunting rituals among northern and southern San groups: Striking similarities. *Africa* 48:117–34.

Lewis-Williams, J. D., G. Blundell, W. Challis, & J. Hampson. 2000. Threads of light: Reexamining a motif in southern African rock art. *South African Archaeological Bulletin* 55:123–36.

Lewis-Williams, J. D., & S. Challis. 2010. Truth in error: An enigmatic nineteenth-century San comment on southern African San rock paintings of 'lions' and 'shields'. *Before Farming*, online journal, http://www.academia.edu/975968/Truth_in_Error_an_enigmatic_nineteenth-century_San_comment_on_southern_African_rock_paintings_of_lions_and_shields_with_D._Lewis-Williams_Before_Farming._9_1article_2._2010.

Lewis-Williams, J. D., & S. Challis. 2011. *Deciphering Ancient Minds: The Mystery of San Bushman Rock Art*. London: Thames & Hudson.

Lewis-Williams, J. D., & T. A. Dowson. 1990. Through the veil: San rock paintings and the rock face. *South African Archaeological Bulletin* 45:5–16.

_____. 2000. *Images of Power: Understanding San Rock Art*. Cape Town: Struik.

Lewis-Williams, J. D., & J. H. N. Loubser. 1986. Deceptive appearances: A critique of southern African rock art studies. *Advances in World Archaeology* 5: 253–89.

Lewis-Williams, J. D., & D. G. Pearce. 2004a. *San Spirituality: Roots, Expressions, and Social Consequences*. Cape Town: Double Storey; Walnut Creek, CA: AltaMira Press.

_____. 2004b. Southern African rock paintings as social intervention: A study of rain-control images. *African Archaeological Review* 21:199–228.

_____. 2005. *Inside the Neolithic Mind: Consciousness, Cosmos, and the Realm of the Gods.* London: Thames & Hudson.

_____. 2009. Constructing spiritual panoramas: Order and chaos in southern African San rock art panels. *South African Humanities* 21:41–61.

_____. 2012. The southern San and the trance dance: A pivotal debate in the interpretation of San rock paintings. *Antiquity* 86:696–706.

Lichtenstein, H. 1812/1928. In A. Plumptre (Ed.), *Travels in Southern Africa in the Years 1803, 1804, 1805, and 1806*. Cape Town: van Riebeeck Society.

_____. 1930. *Travels in Southern Africa in the Years 1803, 1804, 1805, and 1806*. Cape Town: van Riebeeck Society.

Lloyd, L. C. 1889. *A Short Account of Further Bushman Material Collected*. London: David Nutt.

Loubser, J., & G. Laurens. 1994. Depictions of domestic ungulates and shields: Hunter/gatherers and ago-pastoralists in the Caledon River valley area, in *Contested Images: Diversity in Southern African Rock Art Research*, T. A. Dowson & J. D. Lewis-Williams (Eds.). Johannesburg, Witwatersrand University Press, 83–118.

Louw, J. A. 1975. Clicks as loans in Xhosa, in *Bushman and Hottentot Linguistic Studies*, J. W. Snyman (Ed.). Pretoria: University of South Africa, 82–100.

Low, C. 2008. *Khoisan Medicine in History and Practice*. Köln: Rüdiger Köppe Verlag.

Mackenzie, J. 1883. *Day-Dawn in Dark Places*. London: Cassell.

Maggs, T. M. O'C. 1967. A quantitative analysis of the rock art from a sample area in the Western Cape. *South African Journal of Science* 63:100–04.

Malinowski, B. 1926. *Myth in Primitive Psychology*. London: Norton.

Mallen, L. 2008. Rock art and identity in the north-eastern Cape Province, unpublished M.A. thesis, University of the Witwatersrand, Johannesburg.

Marshall, L. 1959. Marriage among !Kung Bushman. *Africa* 29:335–65.

_____. 1960. !Kung Bushman bands. *Africa* 30:325–55.

_____. 1961. Sharing, talking and giving: Relief of social tensions among !Kung Bushmen. *Africa* 31:231–49.

_____. 1962. !Kung Bushman religious beliefs. *Africa* 32:221–51.

_____. 1969. The medicine dance of the !Kung Bushmen. *Africa* 39:347–81.

_____. 1976. *The !Kung of Nyae Nyae*. Cambridge, MA: Harvard University Press.

_____. 1999. *Nyae Nyae !Kung Beliefs and Rites*. Cambridge, MA: Harvard University Press.

Marshall, J., & C. Ritchie. 1984. *Where Are the Ju/wasi of Nyae Nyae?* Cape Town: Centre for African Studies, University of Cape Town.

Marshall Thomas, E. 2003. The lion/Bushman relationship in Nyae nyae in the 1950s: A relationship crafted in the Old Way. *Anthropologica* 45: 73–78.

Marshall Thomas, E. 2006. *The Old Way: A Story of the First People*. New York: Farrar Straus Giroux.

Mazel, A. D. 1992. Changing fortunes: 150 years of San hunter-gatherer history in the Natal Drakensberg, South Africa. *Antiquity* 66:758–67.

_____. 2009a. Images in time: Advances in the dating of Maloti-Drakensberg rock art since the 1970s, in *The Eland's People: New Perspectives in the Rock Art of the Maloti-Drakensberg Bushmen. Essays in Memory of Patricia Vinnicombe*, P. Mitchell & B. W. Smith (Eds.). Pietermaritzburg: University of Natal Press, 81–85, 88–97.

_____. 2009b. Unsettled times: Shaded polychrome paintings and hunter-gatherer history in the southeastern mountains of southern Africa. *Southern African Humanities* 21:85–115.

_____. 2011. Time, colour, and sound: Revisiting the rock art of Didima Gorge, South Africa. *Time and Mind* 4:283–96.

McGranaghan, M. 2012. Foragers on the frontier: The /Xam Bushmen of the Northern Cape, South Africa, in the nineteenth century. Ph.D. thesis, University of Oxford.

_____. 2012/4. In small things consumed: Entomophagy in the Bleek-Lloyd narratives. *Before Farming* 2012/4: article 2.

_____. 2014a. 'He who is a devourer of things': Monstrosity and construction of difference in /Xam Bushman oral literature. *Folklore* 125:1–21.

_____. 2014b. 'Different people' coming together: Representations of alterity in /Xam Bushman (San) narrative. *Critical Arts: South-North Cultural and Media Studies* 28:670–88

_____. In press. 'My name did float along the road': Naming practice and /Xam Bushman identies in nineteenth-century Karoo (South Africa). *African Studies*.

McGranaghan, M., S. Challis, & J. D. Lewis-Williams. 2013. Joseph Millerd Orpen's 'A glimpse into the mythology of the Maluti Bushmen': A contextual introduction and republished text. *Southern African Humanities* 25:137–66.

McGuire, R. H. 1992. *A Marxist Archaeology*. New York: Academic Press.

Mguni, S. 2004. Cultured representation: Understanding 'formlings', an enigmatic motif in the rock-art of Zimbabwe. *Journal of Social Archaeology* 4:181–99.

Mitchell, P. 2002. *The Archaeology of Southern Africa*. New York: Cambridge University Press.

_____. 2009. Hunter-gatherers and farmers: Some implications of 1,800 years of interaction in the Maloti-Drakensberg region of southern Africa. *Senri Ethnological Studies* 73:15–46.

_____. 2010. Making history at Sehonghong: Soai and the last Bushman occupants of his shelter. *Southern African Humanities* 22:149–70.

Mitchell, P., & S. Challis. 2008. A 'first' glimpse into the Maloti Mountains: The diary of James Murray Grant's expedition of 1873–1874. *Southern African Humanities* 20:399–461.

Mohr, E. 1876. *To the Victoria Falls of the Zambesi*. London: Sampson Low, Morton, Searle, and Rivington.

Orpen, J. M. 1874. A glimpse into the mythology of the Maluti Bushmen. *Cape Monthly Magazine* (n.s.) 9(49):1–13.

_____. 1964. *Reminiscences of Life in South Africa from 1846 to the Present Day*. Cape Town: Struik.

Ortner, S. B. 1972. On key symbols. *American Anthropologist* 75:1337–46.

Pager, H. 1971. *Ndedema*. Graz: Akademische Druck.

_____. 1975. *Stone Age Myth and Magic, as Documented in the Rock Paintings of South Africa*. Graz: Akademische Druck.

Paterson, W. 1789. *A Narrative of Four Journeys into the Country of the Hottentots and Caffraria (1777, 8, 9)*. London: J. Johnson.

Pearce, D. G. 2012. Ethnography and history: The significance of social change in interpreting rock art, in *Working with Rock Art: Recording, Presenting, and Understanding Rock Art Using Indigenous Knowledge*, B. Smith, K. Helskog, & D. Morris (Eds.). Johannesburg, University of the Witwatersrand Press, 135–45.

Penn, N. 2005. *The Forgotten Frontier: Colonists and Khoisan on the Cape's Northern Frontier in the 18th Century*. Cape Town: Double Storey.

Philip, J. 1828. *Researches in South Africa*. London: James Duncan.

_____. 1842. South Africa. *Missionary Magazine and Chronicle* 6:183–84.

Potgieter, E. F. 1955. *The Disappearing Bushmen of Lake Chrissie*. Pretoria: Van Schaik.

Prins, F. E., & Lewis, H. 1992. Bushmen as mediators in Nguni cosmology. *Ethnology* 31:133–47.

Propp, V. 1958. *Morphology of the Folktale*. Bloomington, IN: American Folklore Society.

Report and Proceedings of the Government Commission on Native Laws and Customs. Cape of Good Hope. 1883. G.4-'83. Cape Town: Government Printer.

Roberts, A. 1951. *The Mammals of South Africa*. Cape Town: Central News Agency.

Rudner, J., & I. Rudner, 1970. *The Hunter and His Art: A Survey of Rock Art in Southern Africa*. Cape Town: Struik.

Sadr, K. 1997. Kalahari archaeology and the Bushman debate. *Current Anthropology* 38:104–12.

Schapera, I. 1930. *The Khoisan Peoples of South Africa*. London: Routledge & Kegan Paul.

Schmidt, S. 1973. Die Mantis Religiosa in den Glaubensvorstellungen der Khoesan-Völker. *Zeitschrift für Ethnologie* 98(1):102–25.

_____. 1989. *Katalog der Khoisan-Volkserzählungen des südlichen Afrikas* (2 vols.). Hamburg: Helmut Buske Verlag.

Schmidt, S. 1996. The relevance of the Bleek/Lloyd folktales to the general Khoisan traditions, in *Voices from the Past: /Xam Bushmen and the Bleek and Lloyd Collection*, J. Deacon & T. A. Dowson (Eds.). Johannesburg: Witwatersrand University Press: 100–21.

Schoeman, K. 1997. *A Debt of Gratitude: Lucy Lloyd and the 'Bushman Work' of G. W. Stow*. Cape Town: South African Library.

Segal, R. A. 1999. *Theorizing about Myth*. Amherst: University of Massachusetts Press.

Selous, F. C. 1893. *Travels and Adventures in South-East Africa*. London: Rowland Ward.

Shaw, B. 1820. *Memorials of South Africa*. Cape Town, Struik (reprint).

Shennan, S. 2002. *Genes, Memes, and Human History: Darwinian Archaeology and Cultural Evolution*. London: Thames & Hudson.

Shortridge, G. C. 1934. *The Mammals of South West Africa*. London: Heinemann.

Silberbauer, G. B. 1965. *Bushman Survey*. Gaberones: Bechuanaland Government.

_____. 1981. *Hunter and Habitat in the Central Kalahari Desert*. New York: Cambridge University Press.

Skinner, J. D., & C. T. Chimimba. 2005. *The Mammals of the Southern African Subregion*. New York: Cambridge University Press.

Skotnes, P. 1996a. *Miscast: Negotiating the Presence of the Bushmen*. Cape Town: University of Cape Town Press.

_____. 1996b. The thin black line: Diversity and transformation in the Bleek and Lloyd Collection and the paintings of the southern San, in *Voices from the Past: /Xam Bushmen and the Bleek and Lloyd Collection*, J. Deacon & T. A. Dowson (Eds.). Johannesburg; Witwatersrand University Press, 234–44.

Skotnes, P. 2007. *Claim to the Country: The Archive of Wilhelm Bleek and Lucy Lloyd*. Cape Town: Jacana. Athens: Ohio University Press.

_____. 2008. *Unconquerable Spirit: George Stow's History Paintings of the San*. Johannesburg: Jacana.

Smith, B. W. 2010. Envisioning San history: Problems in the reading of history in the rock art of the Maloti-Drakensberg mountains of South Africa. *African Studies* 69:345–59.

Smith, B. W., J. D. Lewis-Williams, G. Blundell, & C. Chippindale. 2000. Archaeology and symbolism in the new South Africa. *Antiquity* 74:467–68.

Smith, B. W., & S. Ouzman. 2004. Taking stock: Identifying Khoenkhoen herder rock art in southern Africa. *Current Anthropology* 45:499–526.

Smits, L. G. A. 1971. The rock paintings of Lesotho, their content and characteristics. *South African Journal of Science Special Publication* 2:14–19.

Solomon, A. C. 1997. The myth of ritual origins? Ethnography, mythology, and interpretation of San rock art. *South African Archaeological Bulletin* 52:3–13.

_____. 2007. Images, words, and worlds: The /Xam testimonies and the rock arts of the southern San, in *Claim to the Country: The Archive of Lucy Lloyd and Wilhelm Bleek*, P. Skotnes (Ed.). Cape Town: Jacana, 149–59.

_____. 2008. Myths, making and consciousness: Differences and dynamics in San rock arts. *Current Anthropology* 49:59–86.

_____. 2009. Broken strings: Interdisciplinarity and /Xam oral literature. *Critical Arts* 23:26–41.

Solway, J. S., & R. B. Lee. 1990. Foragers, genuine or spurious? Situating the Kalahari San in history. *Current Anthropology* 31:109–22.

Sparrman, A. 1789. *A Voyage to the Cape of Good Hope*. London: Lackington.

Spohr, O. H. 1962. *Wilhelm Heinrich Immanuel Bleek: A Bio-Bibliographic Sketch*. Cape Town: University of Cape Town Libraries.

Stanford, W. E. 1910. The statement of Silayi with reference to his life among Bushmen. *Transactions of the Royal Society of South Africa* 1:435–40.

Steedman, A. 1835. *Wanderings and Adventures in the Interior of Southern Africa*. London: Longmans.

Steyn, H. P. 1971. Aspects of the economic life of some nomadic Nharo Bushman groups. *Annals of the South African Museum* 56(6):275–322.

Stow, G. W. 1905. *The Native Races of South Africa: A History of the Intrusion of the Hottentots and Bantu into the Hunting Grounds of the Bushmen, the Aborigines of the Country*. London: Swan Sonnenschein.

Stow, G. W., & D. F. Bleek. 1930. *Rock Paintings in South Africa: From parts of the Eastern Province and Orange Free State*. London: Methuen.

Szalay, M. (Ed.) 2002. *The Moon as Shoe: Drawings of the San*. Zurich: Scheidegger & Spiess.

Tanaka, J. 1969. The ecology and social structure of central Kalahari Bushmen. *Kyoto University African Studies* 3:1–20.

Thackeray, A. I. 1983. Dating the rock art of southern Africa, in *New Approaches to Southern African Rock Art*, J. D. Lewis-Williams (Ed.), South African Archaeological Society Goodwin Series 4:21–26.

Thompson, G., 1827. *Travels and Adventures in Southern Africa*. London: Colburn.

Thompson, S., & W. E. Roberts. 1960. *Types of Indic Oral Tales*. Helsinki: Folklore Fellows Communications, No. 18.

Thornton, R. 2014. The foundation of African studies: Wilhelm Bleek's African linguistics and the first ethnographers of southern Africa, in *The Courage of //kabbo: Celebrating the 100th Anniversary of the Publication of Specimens of Bushman Folklore*, J. Deacon & P. Skotnes (Eds.). Cape Town: UCT Press, 99–117.

Thunberg, C. P. 1796. *Travels in Europe, Africa, and Asia Made between the Years 1770 and 1779*. London: F. & C. Rivington; Cape Town: van Riebeeck Society.

Tindall, H. 1856. *Two Lectures on Great Namaqualand and Its Inhabitants*. Cape Town: Pike.

Traill, A. 2009. *A !Xóõ Dictionary*. Köln: Rüdiger Köppe.

Valiente-Noailles, C. 1993. *The Kua: Life and Soul of the Central Kalahari Bushmen*. Rotterdam: Balkema.

Van Vuuren, H. 1994. Forgotten territory: The oral tradition of the /Xam. *Alternation* 1, 2:57–70.

Vinnicombe, P. 1975. The ritual significance of eland (*Taurotragus oryx*) in the rock art of southern Africa, in *Les religions de la Préhistoire*, E. Anati (Ed.). Capo de Ponte (Brescia): Centro Camuno di Studi Preistorci, 379–400.

_____. 1976. *People of the Eland: Rock Paintings of the Drakensberg Bushmen as a Reflection of Their Life and Thought*. Pietermaritzburg: University of Natal Press.

_____. 2009. Basotho oral knowledge: The last Bushman inhabitants of the Mashai district, Lesotho, in *The Eland's People: New Perspectives in the Rock Art of the Maloti-Drakensberg Bushmen, Essays in Memory of Patricia Vinnicombe*, P. Mitchell & B. Smith (Eds.). Johannesburg: Witwatersrand University Press, 165–90.

Vitebsky, P. 1995. *The Shaman: Voyages of the Soul—Trance, Ecstasy, and Healing from Siberia to the Amazon*. London: Macmillan.

Walker, N., 1996. *The Painted Hills: Rock Art of the Matopos*. Gweru: Mambo Press.

Watson, S. 1991. *Return of the Moon: Versions from the /Xam*. Cape Town: Carrefour Press.

Wells, L. H. 1933. The archaeology of Cathkin Park. *Bantu Studies* 7:113–29.

Weintroub, J. 2013. On biography and archive: Dorothea Bleek and the making of the Bleek Collection. *South African Historical Journal* 65:70–89.

_____. 2014. Colonial adventurer, loyal follower, or problematic afterthought? Revisiting the life and scholarship of Dorothea Bleek, in *The Courage of //kabob: Celebrating the 100th Anniversary of the Publication of Specimens of Bushman Folklore*, J. Deacon & P. Skotnes (Eds.). Cape Town: UCT Press, 151–68.

_____. N.d. *By Small Wagon with Full Tent: Dorothea Bleek's Journey to Kakia, June to August 1913*. Cape Town: LLAREC, University of Cape Town.

Werner, A. 1908. Bushman paintings. *Journal of the Royal African Society* 7:387–93.

Wessels, M. 2010. *Bushman Letters: Interpreting /Xam Narrative*. Johannesburg: Witwatersrand University Press.

_____. 2014. The creation of the eland: A close reading of a Drakensberg San narrative. *Critical Arts* 28:555–68.

Whitelaw, G. 2009. 'Their village is where they kill game': Nguni interactions with the San, in *The Eland's People: New Perspectives in the Rock Art of the Maloti-Drakensberg Bushmen: Essays in Memory of Patricia Vinnicombe*, P. Mitchell & B. Smith (Eds.). Johannesburg: University of the Witwatersrand Press, 139–42, 147–63.

Whitley, D. 1994. By the hunter, for the gatherer: Art, social relations and subsistence in the prehistoric Great Basin. *World Archaeology* 25:356–73.

Whitley, D. S. 2000. *The Art of the Shaman: Rock Art of California*. Salt Lake City: University of Utah Press.

Widdicombe, J. 1891. *Fourteen Years in Basutoland*. London: The Church Printing Company.

Widlok, T. 2004. (Re-)current doubts on hunter-gatherer studies as contemporary history, in *Hunter-Gatherers in History, Archaeology, and Anthropology*, A. Barnard (Ed.). Oxford: Berg.

Wiessner, P. W. 1977. Hxaro: A regional system of reciprocity among the !Kung San for reducing risk. Unpublished Ph.D. thesis, University of Michigan.

_____. 1982. Risk, reciprocity, and socials influence on !Kung San economics, in *Politics and History in Band Societies*, E. Leacock & R. B. Lee (Eds.). New York: Cambridge University Press, 61–84.

Willcox, A. R., 1956. *Rock Paintings of the Drakensberg*. London: Parish.

_____. 1984. *The Drakensbery Bushmen and Their Art*. Winterton: Drakensberg Publications.

Wilmsen, E. N. 1989. *Land Filled with Flies: A Political Economy of the Kalahari*. Chicago: University of Chicago Press.

Wilson. M. 1969. The hunters and herders, in *The Oxford History of South Africa*, vol. 1, M. Wilson & L. Thompson (Eds.). Oxford: Clarendon Press, 40–130.

Wittenberg, H. 2014. The boer and the jackal: Satire and resistence in Khoi orature. *Critical Arts* 28: 503–609.

Woodhouse, H. C., 1984. *When Animals Were People: A–Z of Animals of Southern Africa as the Bushmen Saw and Thought Them and as the Camera Sees Them Today*. Johannesburg: Chris van Rensburg Publications.

Wright, J. B. 1971. *Bushman Raiders of the Drakensberg*. Pietermaritzburg: University of Natal Press.

Wright, J., & A. Mazel. 2007. *Tracks in a Mountain Range: Exploring the History of the uKhalamba-Drakensberg*. Johannesburg: Witwatersrand University Press.

Yates, R., J. Golson, & M. Hall. 1985. Trance performance: The rock art of Boontjieskloof and Sevilla. *South African Archaeological Bulletin* 40:70–80.

Yates, R., J. Parkington, & T. Manhire. 1990. *Pictures from the Past: A History of the Interpretation of Rock Paintings and Engravings of Southern Africa.* Pietermaritzburg: Centaur Publications.

Yellen, J. E., & A. S. Brooks. 1988. The late Stone Age archaeology of the !Kangwas and /Xai /Xai valleys, Ngamiland. *Botswana Notes and Records* 20:5–27.

Yorke, J. 2013. *Into the Woods: A Five-Act Journey into Story*. London: Penguin.

Young, R. B. 1908. *The Life and Work of George William Stow*. London: Longmans, Green, & Co.

INDEX

Compiled by Marthina Mössmer
Numbers in **bold face** refer to illustrations.

!gi: 64–65, 70, 174
!gi:xa 127
  meaning 64
!goïn !goïn 95
!Gurriten-dé 174, 176
!Hāūnu 109, 188
!Hu!hun 79
!Kaha, a.k.a. Jack 194–196
!khauken 65, 110, 174, 190
!Khwa 73, 90, 109, 174
!Kō 57, 73, 126
!Kung 57, 114, 195
!Kwarra-an 197
!Kweíten ta //*kēn* 88
!Nuin-/kúiten 184, 186–188, 190–192, 196, 205
  death 191–192, 197, 207
  meaning of name 190–191
!Xóō 141

/Ai!ae 194–196
/Gao 94
/Han≠kass'o 64, 73, 75, 86, 87, 91, 95, 108, 114, 121, 127, 133, 136, 137, 140, 145, 147, 160, 163, 170, 187–188, 188, 192, 196
  children 79
  father 78
  father-in-law 79
  grandmother 133
  life 78–79
  as narrator 133
  wife 73, 79
/Kaggen 52, 55, 62–63, 110, 114, 128
  and arrows 106
  'makes pictures' 170
  other names 66
  son 83
  'things' 70
  See also Cagn; Mantis
/Kāūnū 189
/*ko:öde* 70
/Kwammang-a 80, 84–85, 87, 91–92, 100–103, 109, 114, 118–120, 146
/Xam 38–40, 67, 95, 96, 103, 107, 110, 112–113, 115, 120, 121, 122, 123, 125, 126, 127, 128, 140, 141, 143, 146, 147, 163, 166, 167, 170, 172, 174–176, 183–192, 194, 196–199, 205, 207, 208
  extinction of 183, 207–209
/Xarraŋ/xarraŋ 39
/Xue 67

//Gauwa 73, 88, 94, 159
  wife 72, 86, 95
//*gaūwasi* 143, 195, 196
//*gwi* 161
//Kabbo **pl. 3**, 68, 75, 78, 80, 84, 87, 91, 96, 101, 108, 110, 111, 114, 117–118, 141, 188–189, 192, 207–209
  daughter 79
  death 98
  life 97–100
  meaning of name 99
  portrait of 35, 97
  as rainmaker 99, 142
  relationship with Mantis 147, 170
'//Kabbo's Song on the Loss of his Tobacco Pouch' 192
//Kanndoro 66, 99

//ke:n 70, 176
//kēn 88
//Kheten 188
//Khwai-hem 83, 84
//Kunn 78–79, 130, 189

≠Gao N!a 72, 95, 139
≠Kágára 109, 188
'≠Kágára and !Hāūnu, who fought each other with lightning' 109
≠Kammi 133
≠Kasin 75, 191

aardvark 141
affines 84, 85, 105, 109, 113, 118, 120, 125, 129, 146, 201
Agama 137, 140
aggression 139
agriculturalists 38, 49, 50–53, 158, 162, 173, 178–180
ales 154
All-Devourer 84
altered states of consciousness 60, 62, 73, 89, 99, 111, 115, 128, 142, 143, 164, 169
  See also dreaming; out-of-body travel; trance
ambiguity 81, 146, 169, 172, 186, 200
ambivalence 63, 89, 131, 142, 200
ancient Greece 7
androgyny 90
animal de passage 90–91
anteaters 141
antelope
  heads 71, 152, 163, 176
  hoofs 152, 163, **164**
  in rock art 154
  skin 121, 122
  wounded 129
Anthing, Louis 207
aprons 102, 112
arguments 146
armpits 65, 135, 145, 149
arrow
  heads 134–135, 136
  links 104, 144
  points 104, 158, 134–135
  poison 93, 104
  shafts 104, 136, 159
arrows 52, 81, 88, 93, 102, 103, 106, 107, 109, 126, 135, 136, 140, 144, **158**, 205
  blunting 134, 135
  deflected 106, 107
  effect of honey on 93
  exchanging 103, 120
  ownership of 120
  parts of 104
  in rock art 157, 158–161
  of sickness 74, 108, 122, 130, 145, 153, 160, 171
artefacts. *See* reversion
assegai 51, 55
avoidance
  relationships 146
  words 114, 129, 168, 196
axes, battle 158

baboon
  eyes 126
  hair 70
bags **pl. 4**, **pl. 5**, 113, 118–119, 121–123, 126, 136, 137, 144, 149
  for arrows 136, 154
  conical 154, 155–157, **156**
  and eland 154, **156**
  quadrilateral 154
  in rock art 152, 154–157, **156**, 164, **165**
  transforming 154–156, **156**
  used for gathering 154
  water 137, 138
Bantu speakers 38, 49, 50, 51, 158, 173, 178–180
Bantu Studies 37, 40–41
barrenness 198
Barrow, John 178
Basotho 162, 167, 174, 178, 179
battle-axes 158
beads 112, 113
beating 110–111

beehives 94
bees 86, 93, 94
'Mother of the ~' 86, 95
nests 68–69, 92
swarming 95
Beetle 136–137, 138, 139, 143, 147
behaviour
antisocial 123, 129
deviant 64
eland 83, 162
belongings. *See* possessions
belts 55, 69, 102
Bible 68
Biesele, Megan **pl. 11,** 44, 76, 98
binary oppositions 45, 200
birds 145, 184, 185–186
Bitterpits 207
Bleek and Lloyd Archive 37, 183
electronic 37
recording 36–37, 98, 99–100, 117, 133, 186–188, 190, 200
Bleek family 39, 40, 42, 47, 48, 49, 66, 67, 79, 97, 98, 99, 207–208
Bleek, Dorothea 36, 37, 40, 65, 78, 84, 98, 99, 107, 110, 117, 118, 124, 129, 134, 137, 142, 183, 192, 207–209
attitude to /Xam 40–41
journey 170, 184
narrative titles 75
Bleek, Edith 40, 98, 99
Bleek, Wilhelm 34–35, **35,** 36, 40, 53–54, 62–63, 64, 97–100, 113, 184, 190, 207–209
death 35, 40, 79
blood 92
eland 81, 101, 154, 167
nasal 66, 71, 109, 121, 122, 151–152, 154, 155, **156,** 159, 163–164, **164,** 188
quagga 118
rain-animal 89
Blue Crane 65
boasting 63, 129
Boers 39, 62–63, 78, 191, 196, 197, 199
killing San people 191, 207
books 208
bored stones 157
Botswana 194
bowmen, in rock art **pl. 6,** 158, 159, 160, 176
bows 52, 102, 103, 111, 126
musical **pl. 7,** 163, **164,** 189, 191, 198
in rock art 154, 157
unstrung 157
bowstrings 134–135, 144–145, 163, 190, 191, 197
breaking 134
materials used for 144
striking 189
Breakwater Prison 40, 79, 97
breathing in 121, 122
bride service 84–85, 105, 109, 146
brushes, paint 168
bull roarer 95
'Bushman' 211
Bushman Dictionary 41, 84, 92, 190
'Bushman rice' 137, 145
buttons 102, 112–113, 127

Cagn 55–57, 59, 62, 64, 66, 69, 69–70, 71, 72, 73, 74, 91–92, 115, 163, 177
cry of 66, 92
See also /Kaggen; Mantis
camp 89, 92, 117, 119
Campbell, John 122
Campbell, Joseph 77, 201
Cape Monthly Magazine 53, **54**
caprice 67
caps 119
eared 174–175
carnivores 87, 128
carrying 154
catch words 44
cats 111, 117, 128, 142
cattle 38, 125, 179, 180, 191, 212
in rock art 164, **165,** 179
Cgoriöinsi 56
Chapman, James 129
charms 55, 56, 57, 66, 69–72, 92, 115, 177
Chopin, Frédéric 202

Christianity 51, 67–68, 76, 90
clapping 194
  in rock art 151, 163
class conflict 204
classless societies 204
claws 125
clicks 11, 51
clouds 130, 140, 163, 187, 189
coals 66, 73, 121
coat of arms **pl. 12,** 35, 208
Cogaz 55–57, 69–70, 72
colonists 38–39, 84, 113, 173, 178, 196, 207
colours, of antelope 92
composition 151
conflict 106–109, **158,** 201–202
  class 204
  resolution 146, 206
  in rock art 157–161, 199–200
  social 115–116, 146
  supernatural 109, 124, 126, 147, 158, 161, 188, 197, 199
Connochaetes gnou 139
conspiring 91, 103, 123
contact 50–53, 178
continuity 104
contradiction 63–64, 206
'Contribution from a Bushman' 63
cosmology 46, 89–90
  Christian 90
  San 166, 169
  tiered 89–90, 141, 144
Coti 83, 162
cousins 84
cracks, in rock face 153, 163
creation myths 51, 80–83, 91, 95–97, 133, 177
  eland 75–77, 152, 161–163, 166–167
  moon 81, 108, 113–115
crie boom 103
Cronus 68
crystals, quartz 113
curing 60, 65, 110

dance circle 61, 115
dancing rattles 161
dangerous creatures 125
dangerous strangers 87, 89, 196
darkness 81, 108–109, 125, 147, 161
Dassie 82–83, 84, 85–86, 119, 128–129, 160
  and bees 86, 95
  and fat 85
dating rock art 212
dawn 95
Deacon, Janette 39
death 108, 115, 124, 140
  harbingers of 141
  and resurrection 114–115
  of trance 72, 115, 177
deviant behaviour 64
Devil 62–63
'Devil's veldschoon' 114
Diä!kwain 33, **38,** 48, 64, 70, 73, 75, 78, 79, 88, 104, 105, 106–107, 108, 112, 121, 122, 127, 129, 140, 159, 164, 176, 179, 183–188, 196, 197, 198, 200
  aunt 190
  death 40
  father 39, 184, 187
  grandmother 39
  life 39–40
  wife 39
diaphragm 151, 159
disorientation 109
dissociation 82
dissonance 105
diviners 52
dogs 125
'doings' 70, 105
domestic animals 125
dots, white 152
Drakensberg. *See* Maloti-Drakensberg
dramatisation 73
dreaming 60, 65, 89, 98, 99, 110, 111, 135, 136, 137, 138, 141–144, 148, 156, 157
drought 198
Drum Dance 61
Durkheim, Emile 205

eagle 55–56, 57, 59, 73, 87, 97
Early Race 81, 88, 103, 136, 140, 141
egalitarianism 69, 94, 204
  subverted 69
eland 64, 66, 90–91, 92, 123, 206
  androgyny of 90
  avoidance words for 168
  bags transforming into 154–155
  behaviour 83, 162
  blood 167–168
  bone 104, 144
  bull 55, 86, 87, 88, 90, 167
  Bull dance 86, 90
  calling 100–101
  cows 167
  creation of 75–77, 80–83, 87–88, 91–92, 93, 95–97, 100, 161–163, 166–167
  dead 126
  death of first 75–77, 80–81, 91–92, 97, 101–102, 105, 167
  dewlap 90, 168
  dying 152, 154–155
  fat 66, 86, 90, 95, 139, 167, 168
  'fight' 106–107
  first 51, 75–77, 101, 162–163, 177
  forelock 179
  gall 81, 102, 108, 114
  gut 144
  hair 179
  heart 86, 167
  hide 87, 104, 144, 167
  and honey 80, 167
  horns 102, 105, 179
  hunting 89–90, 95–96, 129, 148, 179
  kill and trance dance 107–108
  lean 129
  local extinction of 178
  meat 81, 102, 179
  potency 95, 105–109, 152, 168, 177
  respect words for 95
  in rock art **pl. 8**, 149, 152, 154–156, 157, 161–162, 171, 177–178, 179–180
  scent of 96, 107
  sinew 144
  sleeping **156**
  Song 90
  spoor 81, 101, 103, 106
  stomach fat 168
  symbolism 83–84
  and water 88
  wounded 72, 103, 105
elision 114, 174–175, 195
elitism 201
empathy, between hunter and prey 81–82, 103–104, 129
engravings 39, 170
  of eland 84, 90
entrails 134–135, 144–145
Epic of Gilgamesh 76
Equus quagga 140
ethnographic approach 45, 77
euphemisms 93, 95
Europeans. *See* colonists
evil 200
exchange 58, 103, 120, 146
extracorporeal travel. *See* out-of-body travel
eyes 126–127, 141
  Mantis's child's 145
  poking out 118, 126

fables 41
fake paintings 176
familiars, witch 52
farmers 179, 207
  See also agriculturalists; Boers; colonists
fat 85–86, 129
  antelope 125
  as antidote 125
  eating 93
  eland 66, 83, 86, 90, 95, 139, 167, 168
  giraffe 139
  and honey 86–87, 93
  and potency 86
  rubbing with 110, 125
  symbolism of 93
  wildebeest 139

feathers 81, 108, 111, 114, 119, 127–128, 140, 168
fertility 83
fight 88, 92, 97, 99, 101–113, 102, 106, 137, 146, 160
 'the Eland's ~' 106–107
 scenes in rock art 157–161
 See also potency, dangerous concentration of
fingers
 pointing 160, 161
 snapping 160
fire 66, 67, 84, 121, 161
fire sticks 66
First Order 81–82
'firste Bushman' 103
flecks 153, **158,** 159, 160
fluidity 53, 75
flying 111, 127–128
'flying buck' 154
flywhisks, in rock art 152, 157
folklore 41
footprints 120, 187
foraging 138, 143
Fraser, Sir James 36
Freud, Sigmund 7, 77
frogs 88
frontier 40, 173, 207–209

G!ara 66–67, 109
G!kon//'amdima 73
G//aoan 72, 95, 139
G/wi 57, 87
gall
 after death 108
 eland 81, 102, 108, 114
 lion 108
 piercing 102, 108
 as reservoir of potency 108–109
'gallopers' 196–197
garments 112, 113
gemsbok 92, 123
 sinew 144
Ghanzi 180
gifts 146
Gilgamesh 76
Gillen, Francis 36
giraffe
 Dance 94
 fat 139
 Music, owners of 178
 Song 61, 71–72, 94, 177
 urine 73
girls 86, 88–89, 90, 127, 188
'The Girl's Story; the Frogs' Story' 88
gnu 139, 205
god 63, 67, 71, 194
 See also great god; lesser god
god's house 115, 128
gods 7, 64
The Golden Bough: A Study in Magic and Religion 36
good 200
gorge 159
goura 79
gourd 163
grass 134–135
great god 72, 95, 139
 wife of 86, 95
Greece, ancient 7, 68
Guy Fawkes 64

hail 56, 73
hair
 baboon 70
 burning 121
 eland 179
 lions' 125, 187
 potency in 121
 wildebeest tail 139
hallucinogens 60
hare 64, 118, 123
hartebeest 64, 70, 86, 92, 167
 sinew 144
Harvard University 57, 125
headache 102
healers 60, 64, 126

healing 60, 65, 89, 90, 121, 122, 146
health 83
heat 128
herbs, aromatic 70, 71
herders 178
Herero 50
heroes 201
Hoesar, David 37
  See also Diä!kwain
holes 149
  above nape 95
  entering 137, 141, 145
  underground 141, 144–145
  water. *See* waterholes
honey 55–57, 59, 68–69, 78, 80, 85–87, 91–97, 100–101, 119
  being fat 91, 92
  bringing home 91–92, 95
  collecting 93
  combs 55, 59, 93
  Dance 94
  and dancing 95
  drinking 93
  dry 80, 91, 92, 107
  eating 93, 95, 96
  effect on arrows 93
  and fat 86–87, 93
  gathering 86
  hunting 93–94
  hunting observances 93
  and potency 69, 74, 105
  potency of 93–97
  rubbing with 100, 167
  scent of 96
  Song 94
  symbolism of 93
  types of 92
  Western ideas about 68
horns 102, 105, 156, 164, 179
horses 133, 139, 140, 179
  in rock art 164, **165**, 179
'How Sorcerers Assume the Form of a Jackal' 184
'How the Ichneumon Discovered What the Mantis Did with the Honey' 78
How, Marion 167–168, 173–174
hunter-gatherer society 204–206, 212
hunters 72, 82, 89, 93, 103, 123, 128, 129, 139, 204–205
hunting 74, 86, 103, 134–135, 134, 154, 176, 204–205
  ground 89
  honey 93
  observances 88, 93, 95–96, 103–104, 126, 129
  quagga 140
  success 72
  symbolism of 105
hyrax. *See* Dassie

Ichneumon 78, 80, 85, 87, 91–92, 100–103, 117, 118–121, 123–124, 128–130
ideology 59, 204–206
incantations 147
indentured labourers 49
inequality 204–206
initiation 52
in-laws. *See* affines
inquisitiveness 139
interaction 51
  See also contact
intermarriage 51
interpreters 42, 50, 174
intestines. *See* entrails
inversion 201
invisibility 126, 160, 169, 179
iron smelting 51, 158, 159
itching 104

'Jack' 194–196, 196
jackals 184, 185
Jantje Tooren 99
  See also //Kabbo
Jimmy 194–196
John the Baptist 68
joking relationship 100, 129, 146

Ju/'hoansi **pl. 11**, 44, 57, 59, 60–62, 64–65, 66–67, 71–72, 83, 86, 93, 94–96, 98, 99, 103, 107–109, 108, 109, 110–111, 114–115, 116, 120, 125–126, 125, 127, 128, 139, 142, 143, 147, 151, 152, 155, 159, 160, 168, 172, 176–178, 185, 191, 193–196, 197, 204
Judeo-Christian mythology 67–68, 76
Jung, Carl 7, 77
Just So Stories 77

Kalahari 34, 39, 50, 68–69, 98, 104, 119, 125
Kalahari Debate 50–51, 57
Kalahari San 37, 47, 50, 57–62, 71–72, 87, 115–116, 119, 128, 141, 142, 152, 159, 168, 180, 190, 205, 207
kaross 100, 102, 118, 119, 122, 152, 190
  reverting to springbok 122
Kenhardt 97
key symbols 43
Khoekhoe 38, 178, 211, 212
Khoesan 35
Khoisan 212
kia 99, 124, 159, 160
'killing people' 134, 143, 147, 160, 191, 196
Kina-ha 63
kinship 58, 82, 119–120, 145–146, 204, 205, 206
  fictive 58
Kipling, Rudyard 77
Klein Jantje 78
  See also /Han≠kass'o
kloofs 166, 168
kneeboom tree 102
'Kogelman' 137
Korana 84, 207
Koro-tuiten 145
kudu, scent of 96
kukummi 41

labour 204
  symbolic 205
Lake Chrissie 71
Langalibalele 48
leadership 204
legends 41
leopards 125
Lepsius, Karl Richard 34–35, 97
Lesotho 48, 167
lesser god 73, 88, 94, 159
lessons 43, 45, 67, 68
Lévi-Strauss, Claude 7, 42–43, 45, 77, 201–203
Lichtenstein, Heinrich 119
ligament 190
lightning 73, 79, 109, 112, 126–127, 141
lines, in paintings 152–153, 154, 161
Lioness 118–119, 127
lions 87, 109, 112, 113, 124–126, 127–131, 129, 140, 147, 160, 185, 191, 201
  claws 125
  ears 125
  eating people 191
  footprints 120
  fur 125
  gall 108
  of god 126
  hair 125, 187
  mice resembling 141
  paws 187
  potency 108
  respect word for 187
  and shamans 124–126
  spoor 118, 120, 187
  transformation into 108, 125–126, 186–187, 190, 191, 196
  white settlers as 207
  wounded 126
Little Lion 118–119, 122, 123–124, 126–127
Lizard 137–138, 140, 143, 144, 146, 147
  and rain 140
Lizard's song 137
Lloyd, Lucy 34–35, **36**, 37, 41, 64, 72, 78, 79, 84, 97–100, 103–104, 107, 110, 111, 112, 117, 118, 122–123, 127, 133, 137, 160, 183, 184, 186–188, 186, 190, 192–193, 196, 207–209
locusts 68, 70, 74, 145, 174, 189
loneliness 198, 208

'Lord of the Animals' 65
louse 64, 104

magic 70, 105, 121, 187
'magic expedition' 110
magician 99, 186
'make water' 72
Malinowski, Bronislaw 77, 203
Maloti expedition 48–50, 53
Maloti San 50, 64, 83, 139, 162–163, 167, 177–178, 179
Maloti-Drakensberg **pl. 2,** 48, 52, 59, 153, 158, 166, 168, 178
mamba song 71
Mantis 47, 53–54, 55, 59, 62–67, 74, 78, 80–93, 99, 100–125, 126–131, 135, 147–148, 162–163, 167, 169–170, 177, 179, 200, 203, 206
  ability to change form 64, 179
  adopted daughter 84
  affection for eland 148
  affines 85, 87, 105, 120, 125, 129, 201
  antisocial behaviour of 123, 129
  as bull eland 87
  call of 66, 91–92, 100–101, 166
  child's eye 145
  dreams 136, 137, 138, 141, 149
  family 82, **83,** 91, 146
  fight with Meerkats 97, 101–113, 143
  fight with the Cat 111
  grandson 80, 85
  grows feathers 111, 119, 127–128
  guises 64, 104, 123
  as a hare 123
  as hunter 86
  as insect 127–128
  and Lions 117–124, 127–131
  as a louse 104
  names 99
  prayer to 139, 205
  as shaman 201–202
  as a snake 71, 179
  son-in-law 80, 84–85
  wife. *See* Coti; Dassie
  See also /Kaggen; Cagn
Mantis religiosa 127
mantis, praying 63, 64, 66, 127–128, 147, 174
The Mantis and His Friends 77, 134
'The Mantis and the Cat' 111, 117, 128, 142
'The Mantis, the Lizard, and the Mice' 137–138
'The Mantis, the Mice, and the Beetle' 136–137
'The Mantis and the Tortoise' 117
map **pl. 1,** 184
Mapote **pl. 9,** 167–168, 173–174, 176–178
marriage 51, 146
  forced 105, 109
  rites 86, 90
Marshall family 57, 61
Marshall, Lorna 184–185
Marshall Thomas, Elizabeth 125
material culture 51
Mauss, Marcel 205
meaning 43
  indigenous 44
meat 83
  distribution of 120
  eland 81, 83, 179
  'infected' 179
  quagga 129, 140
  sharing 59
  springbok 138
medicine songs 60, 71–72, 94, 142–143, 144, 151, 177
  and potency 151
medicines 93, 121, 161
Meerkats 80, 81, 85, 87, 91, 92, 97, 99, 102–110, 123, 141, 143, 201, 207
  buttons 112–113, 127
Melissa 68
men's tasks 86
metaphor 93, 128
Mice
  Long-Nosed 134–138, 140–141, 146
  resembling lions 141
  Striped 135–138, 140–141, 145, 146, 147
milk 89
missionaries 51

modesty 129, 205
mongoose 8
monsters 83, 84, 98
moon
  creation of 81, 108, 113–115
  full 114
morality 59, 67, 130
morals 45, 68, 76
'Mother of the Bees' 86, 95
motifs 42
Müller, Max 113
music 44, 71–72, 177, 194
musical instruments **pl. 7**, **pl. 10**, 79, 163, 164, 189, 191, 194, 198
myth, theories of 77, 113
mythemes 7, 42–43, 45
mythology 7, 42, 67
  Judeo-Christian 67–68, 76
  San 169
myths 41–42, 201–206
  ancient Greek 7, 68
  canonical 202
  function of 201–206
  hero 201
  interpretation of 76, 195
  lessons 43, 45, 67, 68
  link with rock art 149, 150
  meaning 43, 67
  morals 45, 68, 76
  Sumerian 76
  See also creation myths

n/om 65, 94, 121, 159, 160
  owners of 177
n/om k"xausi 65, 157
Nama 211
names 99
narration 73, 99–100, 117
narratives 41–45
  complete 75–77, 193
  retelling 117
  titles of 75, 136
nasal blood 66, 122, 159, 188
  in rock art 151–152, 154, 155, **156**, 159, 163–164, **164**
Native Races of South Africa 49
needles 160
new maidens. *See* puberty, girls'
Nguni 50, 51, 52
  diviners 52
Nharo 57, 93, 146, 160, 180, 205, 207
Nkonyana 179
nose
  bleeding from 66, 71, 109, 121, 122, 123, 151–152, 154, 155, **156**, 159, 163–164, **164**, **165**, 185
  'Little' 179
  possessing 188
  as seat of potency 188
notebooks. *See* Bleek and Lloyd Archive
novices 159, 160
nuggets 42–46, 57, 65, 66, 68, 69, 72, 73, 76, 77, 80, 87, 88, 106, 107, 109, 110, 121, 123, 145, 150–154, 158, 160, 170–172, 174, 194, 195, 202–204

ochre 168
omens 80
onomatopoeia 196
opposites 200
Orange Free State Monthly Magazine 63
Orpen, Joseph Millard 48–50, **48**, 53–55, 65–66, 71, 72, 75, 83, 91–92, 99, 104, 167, 173–174, 178
ostrich
  eggshell beads 112, 113
  eggshells 118, 184
  feathers 81, 108, 114, 134, 135
  hunting 176
out-of-body travel 65, 72, 110, 126, 128, 141, 143, 159, 176, 185–186, 194, 196, 205
Ovambo 50
ownership 68–69, 96, 103, 143

paint 167–168
  brushes 139, 168
  flecks 153, **158**, 159, 160

painters 169
paintings **pl. 6, pl. 9,** 39, 49, 50, 52, 53, 61, 139, 149–172, 179
  of /Kaggen 64
  of bees 93
  of bleeding from the nose 66
  of eared caps 174–175
  of eland 83, 177–178, 179–180
  fake 176
  of honeycombs 93
  of humans with antelope heads 71
  of karosses 123
  red lines in 152–153, 154
  of transformed shamans 123
  shaded polychrome 179–180
pastoralists 38, 84, 178, 212
pawed creatures 87, 100, 105, 126
pegs 164
People of the Eland 174
perfit 64
performance 100
perspiration. *See* sweat
Phuthi 49, 179
pigment 168
plenty 83
poetry 183, 192–193
pointing 160, 161
poison 93, 104
pool 71
Porcupine 73, 84–85, 87, 105, 119, 128–129
  and fat 85
  father 84
possessions 81, 96, 97, 102, 103, 109, 110, 113, 119, 127, 143, 163
postures
  arms backward 151, 152, 154, 163, 179
  bending forward 151, 164
  raised knee 160
potency 60, 64, 65, 66, 70–71, 88, 90, 93–97, 114, 144, 154–156, 159–161, 174, 176, 177–178
  asking god for 151
  of bees 93
  boiling 67, 159
  burning 160
  dangerous 127
  dangerous concentration of 88, 92, 107, 121, 160
  eland 95, 104, 105–109, 152, 168, 177
  of fat 86
  in hair 121
  harnessing 107
  of honey 69, 74, 93–97, 105
  protective 177–178
  of rain 113, 127
  in rock art 153
  and scent 121–122
  seat of 188
  of songs 194
  in sweat 121, 145
  of urine 73
  See also !gi:; n/om
pots 167
praying mantis 63, 64, 127, 147
predators 89
  See also carnivores; lions
prestige 93–94, 205, 206
Primal Time 81–82
prison. *See* Breakwater Prison
processions, in rock art 151
profeet 64
Prometheus 66
pronunciation 11
property 68–69
prophet 64
puberty, girls' 86, 88–89, 90, 127, 188

qhang qhang 168
Qing 47, 49–50, 51, 53–55, 59, 65–66, 71, 72, 75, 91–92, 104, 115, 162–163, 166, 167, 173–174, 177, 179
quagga 92, 118, 134–136, 140, 141
  blood 118
  hunting 140
  leading hunters to lions 140
  meat 119–120, 129, 140
  sinew 144
  tail 140

quagga, cont.
  and wildebeest 139
quartz crystals 113
quivers 84, 101, 102, 106, 119, 136
  reverting to springbok 121
  in rock art 158
Qüuisi 55, 56, 69, 163
Qwanciqutshaa 64, 163

rain 72, 73, 83, 86, 89–90, 109, 113, 124, 127, 136, 140, 166, 174, 187–189, 205
  angry 122
  doctors 52
  male 73, 189
  Song 90
rain-animals 73, 89–90, **164**, 187–189, 191, 197
  horns 164
  in rock art 152, 154, 163–166, **165**
rainmakers 39, 78–79, 98, 99, 109, 129, 163, 180, 184, 186, 187–189
rainmaking 52, 65, 67, 78–79, 89, 127, 142, 152, 191
  bow 198
  failure of 189
  in rock art 164
  songs 187, 191, 192, 196, 198, 205
  techniques 191
'rain's legs' 187
'rain's things' 71, 88
rattles, dancing 161
raw material 114, 121
raw materials. *See* reversion
realms 89
reconstitute 167
recording 36–37, 98, 99–100, 117, 186–188, 190, 200
  process of 133
reeds 88, 101, 104, 136, 162, 163, 168
relationships 58–59
  avoidance 146
  exchange 58
  joking 85, 100, 129, 146
  personal 57
  sharing 57, 59
  social 69
  See also kinship
religion 59–68
repetition 100, 107
resources 204
respect words 95, 114, 187, 196
resurrection 114–115, 137, 167
  of eland 105
retelling 117
reversion, of artefacts to raw materials 87–88, 104, 114, 121, 122, 167
rhebuck
  heads, men with 163
  in rock art 154
rhetorical questions 196
ribaldry 66
rites of passage 90–91
ritual 60–61, 82
  boys' first-kill 90
  girls' puberty 86, 88–89, 90, 127, 188
  hunting 88, 93, 95–96, 103–104, 126, 129
  knowledge 204
  marriage 86, 90
  specialists. *See* shamans
riverbed 136–137
rivers 163
rock art 39, 62, 93, 149–172, 205, 208
  dating 212
  link with myth 150
  See also engravings; paintings
rock engravings. *See* engravings
rock face **pl. 8**, 153, 154–155, 158, 161, 163–166
rock paintings. *See* paintings
rock rabbit. *See* Dassie
rock shelters 117, 164, **165**, 168
rubbing
  with coals 121
  with fat 110, 125
  with honey 80, 92, 100, 167
  with nasal blood 121
  with sweat 65, 145

San
  cosmology 89–90
  'creation' of 177
  egalitarianism 69, 94, 204
  genocide 183, 184, 207–209
  Kalahari. *See* Kalahari San
  Maloti. *See* Maloti San
  morality 59, 67
  names 99, 190
  northern 39
  origin of term 211
  plight of 207–209
  religion 177, 203, 206
  romanticized view of 99
  society 203, 204
  southern 39, 47, 52, 58, 61, 152, 159, 183, 207
  See also Ju/'hoansi; /Xam
sandals 80, 144
  See also shoes
Satan 62–63
scent 137
  of eland 96, 107
  of honey 96
  of kudu 96
  of nasal blood 121, 159
  and potency 121–122
  of whirlwind 121
  of wildebeest fat 139
Schmidt, Sigrid 42
Schumann, Robert 202
scolding 129, 130
scratching 104
screens 134, 135
sexual
  allusions 66, 105
  intercourse 93, 95
shaded polychromes 179–180
shamanism 62, 65, 116, 160, 188
shamans 52, 60, 62, 64–65, 69, 72, 74, 86, 91, 93, 95, 99, 107–109, 110–111, 115, 116, 121–126, 128, 130–131, 133, 141–144, 145, 146, 147–148, 151, 154, 169, 170, 171, 176–177, 180, 184, 185, 201, 204–206
  benevolent 124, 126
  dead 125, 170, 187–188
  death of 70
  of eland 176
  eyes 127
  fighting 109
  of the game 65, 187–188
  and karosses 122–123, 152
  killing enemies 143
  and lions 124–126, 127, 186–187, 190
  and locusts 70
  of locusts 174, 189
  malevolent 124, 126, 130, 143, 145, 191, 196–200
  of mantises 174
  novice 159, 160
  pain experienced by 125, 190
  and potency 159–161
  protecting people 143–144, 147, 161, 177, 196
  of the rain 98, 140, 152, 164, **165**, 174, 186, 187–189
  respected 204
  in rock art 152
  social position of 204–205
  of springbok 147, 174, 176
  transformed 152, 185–186, 196
shape shifting. *See* transformation
sharing 57, 59, 69, 118
  etiquette 59
  meat 59, 119–120
sheep 179
shepherd 179
shields, in rock art 158
shining 108, 127, 141
shiny objects 113, 127
shivering 111, 159
shoes 80, 81, 87–88, 101, 104, 114, 119, 167
  See also sandals
shrews 140
sickness 60, 122, 177, 205
  expelling 95
  in rock art 153

sickness, cont.
  sniffing out 122
Silayi 52
sinew 144, 190
singing 128, 194
skin
  animal 121, 122
  lizard 140
  snake 71, 140
sky 89, 103, 108, 111, 113, 115, 119, 128, 144, 163, 188
slavery 49
smoke 67, 73, 86, 121, 139
snake
  medicine 71
  powder 71
  skins 71, 140
snakes 64, 71, 141, 167, 179
  with antelope heads 152, 166
  in rock art 152, 166
  transformation into 188
sneezing 109, 122
sniffing 122
Snore-white-lying 122
snoring 110, 122–123
social
  classes 204
  conditions 201, 204–206
  pressure 173
  relations 69, 119, 130, 133, 143, 146, 148, 180, 205
  status 116
  tensions 92, 116, 121, 123, 146, 148, 180, 189
'Song of the Broken String' 32, 46, 183, 191, 192–193, 195–200
songs 115, 192, 193–196
  medicine 60, 61, 71, 191, 194
  personal 191, 192, 195
  rainmaking 187, 196, 198, 205
  types of 191, 192
sorcerers 64, 184, 185, 186, 188, 197
sorcery 70
Sotho 50
sparkling 112
Sparrman, Anders 207
spears 51, 126, 137
Specimens of Bushman Folklore 35, 37, 97, 183
spelling 12, 190
Spencer, Sir Walter 36
spine 159, 160
spirit realm 62, 90, 111, 121, 122, 124, 130, 143, 150, 151, 153, 157, 159, 160, 162, 163–166, 163, 171, 174, 177, 188, 192, 194, 197, 198, 203, 205
spirits
  contacting 96
  of the dead 88, 89, 108, 111, 115, 125, 130, 143, 152, 160, 187–188, 191, 195, 196–200, 198
  malevolent 60, 143, 145, 147, 148, 153, 177, 191, 196–200
  marauding 143, 161, 166
  protection against 143–144, 177
  summoning 157
  'who walk at night' 114
springbok 92, 122, 147, 161, 174, 176, 189
  cap 174
  hunting 137
  meat 138
  sinew 144
  skin 121
  wounded 137
springs 139
'Ssū-!kúï-ten-ttā 122
stalking 176
Stanford, Walter 52
status
  changes in 90
  social 116
  See also prestige
status quo 203
steps, in rock face 153, 163, **164**
stereotyping 155
sticks 56, 57, 69, 122, 134, 135, 138, 146, 167, 179
  dancing 151
  digging 157

fire 66
of medicine 161
stomach muscles 160, 164
stones, bored 157
Stow, George 49, 61, 176
strangers 87, 89, 196
strength 83
string 190–191, 197–199
structuralism 45, 76–77, 200
subservience 203
supernatural potency. *See* potency
supernatural realm. *See* spirit realm
superpositioning, in rock art 155
factitious 157
surreal realms 172
sweat
from armpits 65, 145
and potency 121, 145
rubbing with 145
synecdoche 43–44, 153

tails 133, 134–135, 136, 139, 140
talismans 70
T'ang 63
Tānō !khauken 174, 190
tassels 123
telephone 72
tension
between affines 115, 118, 120, 146
social 92, 116, 121, 123, 146, 148, 171, 180, 189
termite larvae 137, 145
therianthropes, in rock art 152, 176
Thikoloshe 52
thongs 164, 174, 188, 190–191, 198
thorns 159, 160
threads 190
'threads of light' 72, 127, 152–153, 197, 198
thumb-piano **pl. 10**, 194
thunder 109, 188, 196–197
thunderbolts 126–127, 141
See also lightning
thunderstorms 72, 73, 126, 189
ticks 142
tiered cosmos 89, 89–90, 141, 144
tinderbox 66, 99
titles 75, 136
Tixo 63
T'koo 63
Tokoloshe. *See* Thikoloshe
tortoise shells 66–67, 73, 86, 139
trampling 135
trance 52, 66, 72, 86, 89, 99, 110, 111, 121–123, 121, 125–126, 128, 142–143, 151, 152–153, 155, 159, 169, 185, 197
disorientation during 109
'dying' during 72, 115, 124, 177
and music 194
physical effects of 128
in rock art 154
'trance buck' 154
trance dance 60–62, 65, 66, 71, 82, 86, 93, 94, 95, 96, 110–111, 114–116, 128, 139, 142, 144, 146, 147, 148, 152, 157, 168, 169, 171, 176, 177, 179, 194, 205, 206
after eland kill 107–108
evidence of 61
fire at 67
at full moon 114
in rock art 151, 163–164
trance postures. *See* postures
transformation 64, 88–89, 94, 95, 122, 123, 139, 140, 144, 152, 169, 184, 185–186
into animals 67
into birds 185–186
into insects 127–128
into a jackal 185
into lions 108, 124–126, 125–126, 191
of objects 87–88, 104, 114, 121, 122, 167
into a snake 188
into a tree 127
transportation, magical 143
travel
out-of-body 65, 110, 126, 128, 141, 143, 159, 176, 185–186, 194, 196, 205
real-world 114, 184–186, 189, 208
underground 66, 74

trees 81, 102, 103, 110, 111, 127, 140
trembling 65, 102, 110–111, 115, 125, 143, 174, 190
trickster 63, 64, 66–67, 68, 93, 107, 109, 114, 115, 118, 128, 129, 139, 148, 169
    ridiculing 73–74
Tswana 50
tusks 155–156

uKhahlamba 48
    See also Maloti-Drakensberg
underground 66, 68, 141, 144–145, 187
    travel 66
underwater, as trance metaphor 128, 163
underworld 144
Unkonagnana 179
urine 72–73

van der Post, Laurens 99
Vanwyksvlei 78
versification 192–193
vertebral artery 159
Vinnicombe, Patricia 174
vision, affected during trance 128
'A Visit to the Lion's House' 117–119, 130
visitors 119

water 90, 118, 128, 166, 174, 184
    bag 137, 138
    bull 164
    descent into 128
    and eland 88
waterholes 65, 68, 71, 80, 88–92, 91–92, 100–101, 107, 118–119, 128, 144, 162, 163, 164, 166, 167
    bitter 189, 207
wealth 94
weightlessness 128
whirlwind 88–89, 121, 140, 188
    scent of 121
whispering 91, 103, 118, 123
'Why the Wildebeest Has a Light Tail' 133, 134
wildebeest 133–136, 138, 139–140, 141, 143, 144–145
    black 133, 139
    entrails 134–135, 144–145
    fat 139
    hair 139
    and quagga 139
    in rock art 149
    skull 139
    tail 133, 139
'The Wildebeest, the Mice, the Quaggas and the Mantis' 133
wind 88
wings 128
winking 123–124
witch doctors 52
witch familiars 52
women
    clapping 151
    Drum Dance 61
    as shamans 60
    at trance dances 161

Xa-ttin 39, 184, 187–188, 191–192, 197–198, 205
Xhosa 63
Xué 114

zam 139
    See also tortoise shells
zebra 140
Zeus 68

ABOUT THE AUTHOR

David Lewis-Williams is professor emeritus at the University of the Witwatersrand in Johannesburg, South Africa. He founded and was former director of the highly regarded Rock Art Research Institute at Wits University. He is internationally known for his ground-breaking work on the art and beliefs of the southern African San, the Upper Palaeolithic art and Neolithic monuments of western Europe, ancient shamanism, and the neuropsychology of religious experiences. Author of over 120 articles and 19 books on these topics, he has been honored by the American Historical Association, the Society for American Archaeology, and the Royal Anthropological Institute. He was awarded honorary doctorates by the universities of Cape Town and the Witwatersrand. In 2000 he was invited to translate the postapartheid South African national motto into the extinct /Xam San language.